EX LIBRIS

G. HOUGH

J.R. Hudson

Fifty Favourite Nymphs

T. Donald Overfield

Fifty Favourite Nymphs

with illustrations by the author

Ernest Benn Limited

Published by
Ernest Benn Limited
25 New Street Square, London, EC4A 3JA
& Sovereign Way, Tonbridge, Kent, TN9 1RW

First Published 1978
© T. Donald Overfield 1978
ISBN 0-510-22516-0
Printed in Great Britain

Contents

Contents

Foreword

Flytying must be one of the fastest growing crafts—I refuse to call it an art—in this country. The regular opening up of new stillwater fisheries and the mounting pressures on the available streams and rivers indicate an explosion of interest in flyfishing. Many newcomers to the sport of fishing the fly then desire to take a trout on a fly they have tied and so another fur and feather manipulator is born.

This book, *Fifty Favourite Nymphs*, the first of a series that will eventually cover dry flies, wet flies, lures, salmon flies, etc., is intended to bring to the expert and novice flydresser alike those patterns that have qualified for the term 'favourite', a word that can be applied equally to the ancient Gold Ribbed Hare's Ear or such relatively modern patterns as Richard Walker's Chomper series. The word 'favourite' in the title is, therefore, to be interpreted variously. Any one nymph may be included here as a favourite of its designer, a favourite of countless anglers or just a personal favourite of the author. The reader may care to guess in which particular category each fly belongs. If your own personal favourite is missing I seek your indulgence. Perhaps it will be included in a subsequent volume.

It is also important to mention that the term 'nymph' has been applied loosely, and that I include some shrimps and beetles, for today many flyfishers are very interested in fishing such artificials in the manner of the nymph.

With certain exceptions I have illustrated the build-up of the artificial as I would tie it for trouting, not for exhibition work. Some flytyers may criticize my methods, especially my way of tying in the hackle after the whisks, body, ribbing and thorax have been completed. I know that many prefer to tie in the hackle as one of the first stages in the tying of the fly. I would not argue with them for I also use this method when neatness is uppermost in my mind. However, in defence of my usual method I would say that nothing is more infuriating than almost to complete a pattern and then when winding the hackle discover that it is one of those that knows not its father and refuses to wind

7

in a satisfactory manner. In any case I doubt if the final judge, the trout, really cares.

In my drawings I have attempted to show the basic structure of these nymphs, most illustrations being drawn from life, for many of the flydressers mentioned in the book kindly provided me with patterns of their work. I do hope that you will find as much pleasure tying them and fishing with them as I have had in painting them.

You can be sure they are all true favourites, accepted trout takers wherever the rainbows and the brownies come to the nymph.

T. Donald Overfield
Solihull
Warwickshire, 1977

Fifty Favourite Nymphs

Silver Corixa

Price

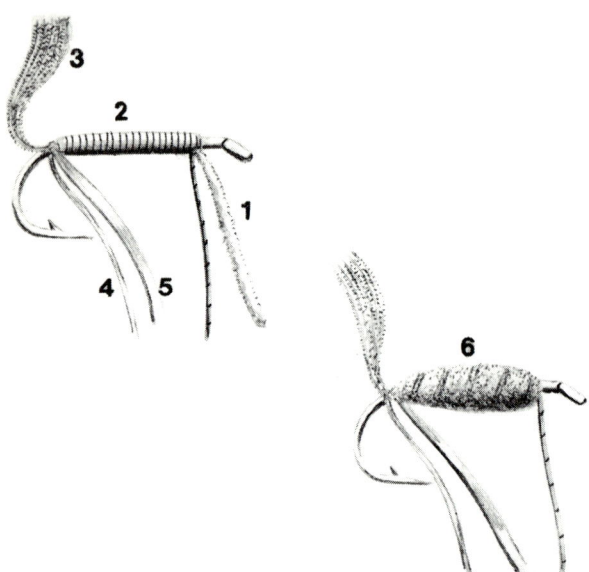

The name of Taff Price is well known to an ever growing band of anglers, for not only is he a most inventive flytyer but also a competent amateur entomologist. His book *Rough Stream Trout Flies* is an authoritative work, but he is equally at home on still waters, as this pattern shows.

It is my firm conviction that really good artificials are usually the result of first-hand study of the natural counterpart and nowhere is this better demonstrated than in Price's version of the corixa. He spent countless hours watching the behaviour of the inhabitants of his research tanks and noticed that this particular species appeared to have the distinguishing habit of shooting up to the surface, where it gathered a bubble of air on its underside, then promptly shooting back to the depths where it proceeded to spread the trapped air on the unwettable hairs over its abdomen. During its rapid journeys to and from the surface it looked remarkably like a bubble of quicksilver, hence the overall silver body of the artificial.

The hook size is between 10 and 14. Start the tying silk,

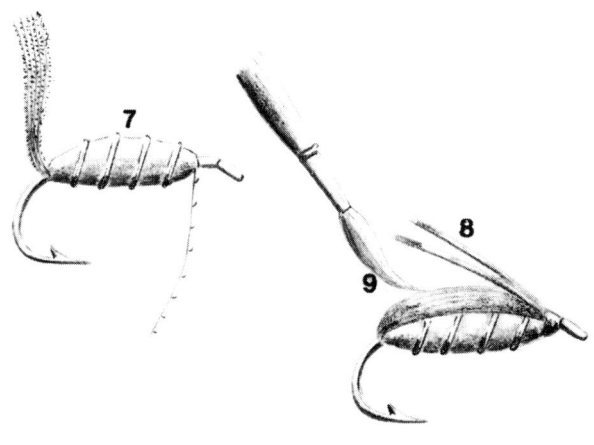

colour optional, down the hook shank, catching in a length of wool, 1. Again the colour is optional for it only forms the underbody. Continue the silk to the bend, 2, then tie in a bunch of cock pheasant tail feather fibres, 3, and a length of oval silver tinsel, 4, followed by a length of flat silver tinsel, 5.

Wind the wool down the shank, backwards and forwards, aiming at a plump and tapered shape, 6. Remove waste material. Wind the flat silver tinsel in even overlapping turns up the body, followed by evenly spaced ribs of the oval silver tinsel, 7. Tie in and cut off waste ends.

Pull the cock pheasant tail fibres over the body from tail to head and secure, removing waste ends. Tie in two cock pheasant tail fibres as paddles, 8. Complete the head with a whip-finish. Varnish the paddles and the back, 9, for added durability.

The pattern can be fished throughout the season. It is very effective as a dropper when fishing on the point a lure that calls for a rapid retrieve, for the corixa is a fast mover.

White Nymph

Nice

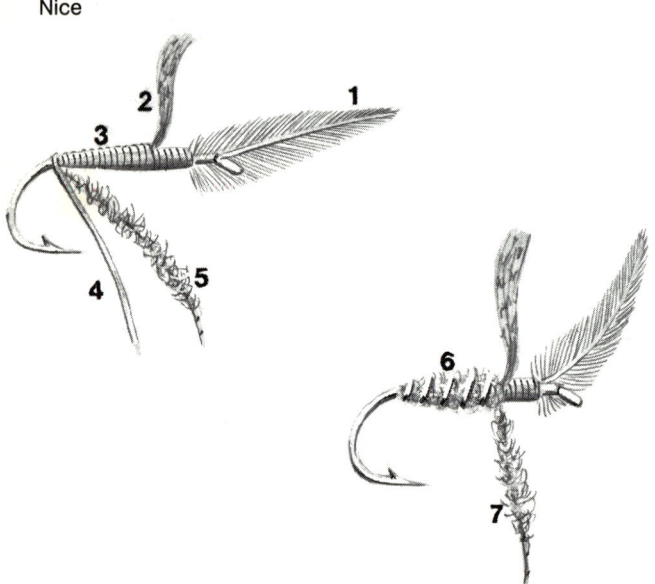

It has been my pleasure to examine artificial flies tied by many of the craft's top practitioners, contemporary and of past eras, but I can say with fair certainty that for incredible neatness and skill in the manipulation of fur and feather James Nice, of Sidmouth, Devon, must come very close to the top of the list. He has not been responsible for a great number of new patterns, believing that the old flies have many years of life left in them yet, but his White Nymph has become a strong favourite of those stillwater flyfishers who help to form the small but highly critical coterie of his regular customers.

You will get some idea of Nice's fanaticism for detail when you note that he even states the number of turns of tying silk between operations, based of course on standard hook sizes.

The White Nymph is tied on a size 12 hook. The silk, of any pale colour, is wound for six turns down the hook shank at which point a pale ginger cock hackle is tied in, 1. The silk is continued for a further twelve turns when the

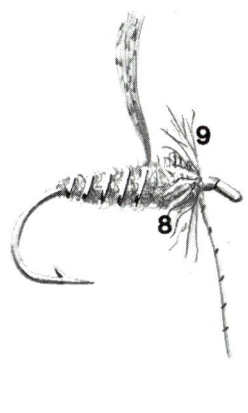

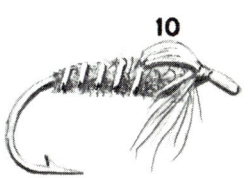

wing cases of any pale mottled feather are tied in on top of the shank. The waste ends are trimmed to a taper and are trapped under the turns of silk as you continue down the hook, 2 and 3. At the bend tie in the ribbing material, 4, the thinnest silver tinsel. Nice insists that it must not be lurex. Now wax the tying silk and dub with white DFM floss fibres, 5. Wind the dubbed silk up the shank to form a tapered body and rib with five turns of tinsel, 6.

Re-dub the silk with white DFM floss, 7, and form a pronounced thorax, 8. Wind the hackle, two turns only, 9, and pull the fibres downwards and rearwards. Bring the wing case material over the thorax and bind down, 10. Clip off waste and complete the nymph with a whip-finished head, then varnish with clear varnish.

You may not be able to recreate Nice's incredible neatness but you can bring to life a most effective pattern that has found great favour on the stillwater fisheries of the West country, and also on quite a few chalk streams if tied down to size 14, or even size 16.

Mayfly Nymph

Walker

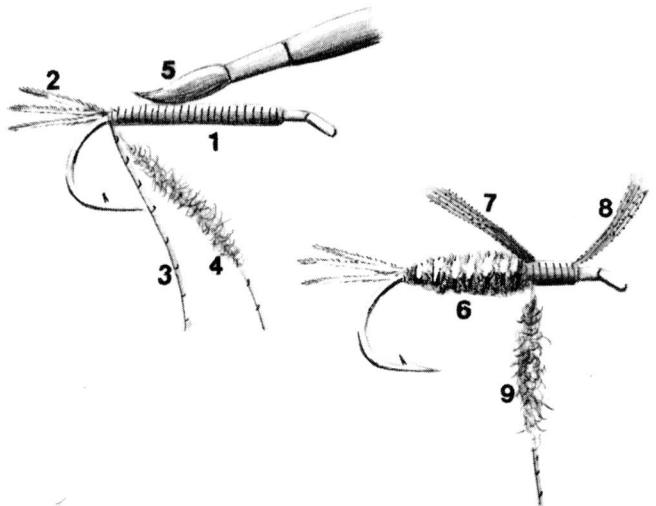

Richard Walker has long been recognized as a most experienced all-round angler, though in the context of this book we are concerned with his contribution to the craft of flyfishing, and more particularly flytying. To cover all the patterns devised by Walker would call for a book in itself and so I have had to be somewhat selective in my choice of patterns, with his Mayfly Nymph being the first of that small number. I am in the fortunate position of being able to fish a stream that still has a hatch of mayfly, so I can recommend this artificial. Richard Walker is equally certain that this pattern works very well where mayfly naturals no longer exist!

The hook is the usual long shank variety. Take the tying silk down the shank to the bend, 1. Now tie in three fibres of natural pheasant tail, 2, followed by a length of brown nylon thread, 3. Re-wax the tying silk and dub with shredded Angora knitting wool or plain sheep's wool, both of a very pale yellowish-buff colour, 4.

Now coat the tying silk on the hook shank with varnish,

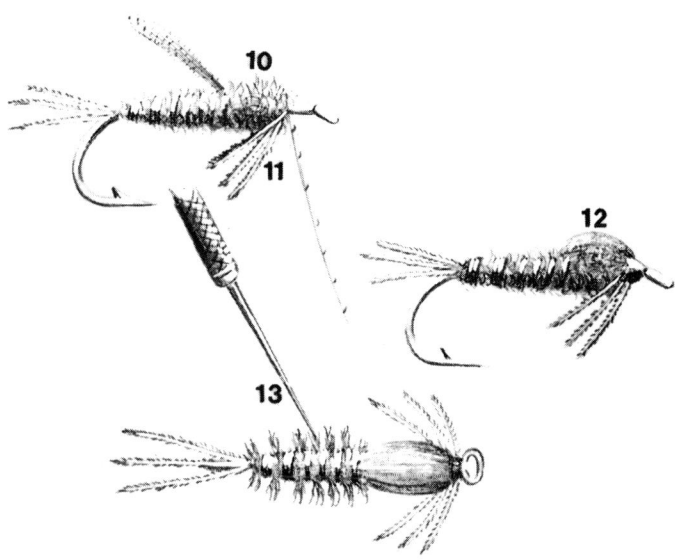

5, following quickly with a wound body of the dubbed silk, 6. This is followed by ribs of the brown nylon thread, making the first turns of silk nearest to the tail thicker than the other turns; say four to five touching turns compared to the remainder. Strip off the remaining dubbing from the tying silk and tie in on top of the shank a bunch of pheasant tail fibres, 7. Wind the silk forward and leave the rest of the fibres pointing out over the eye, 8. Return the silk to the edge of the body material and re-dub with the same dubbing, 9. Wind on as the thorax but do not make too thick, 10. Bring the pheasant tail herls that are pointing forwards down on either side of the shank and pull rearwards to form the legs, 11. Secure with the tying silk. Now bring the rearward facing pheasant tail fibres over the top of the thorax to create the wing cases, 12. Finish off the head with a neat taper and varnish. When dry take a dubbing needle and gently tease out the fibres of the body material along the side, 13. (N.B. An underbody of lead foil may be added for deeper fishing.)

Torp's Reed Smut Nymph

Jacobsen

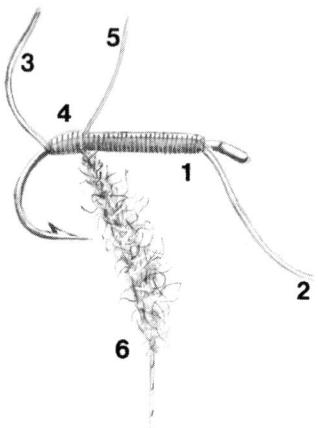

The name of Preben Torp Jacobsen of Denmark is not un-
familiar to knowledgeable British flyfishers for he has con-
tributed many interesting and thought-provoking articles
on flytying and entomology to the magazine *Trout and
Salmon* over many years. His occasional visits to this
country have brought him many friends and admirers
through first-hand knowledge of his undoubted skills and
his highly amusing personality. For more years than I care
to remember I have corresponded with Jacobsen, generally
on the subject of flytying techniques, and I can vouch for
the fact that he is one of the world's most skilled practitioners; a view which I am sure is shared by many others.

It is flyfishing's loss that no British publisher has yet
translated his two books *Tørflue Fiskeri*, published in 1965
and *Nymfe Fiskeri* of 1972, for they are classics. You may
consider this praise rather fulsome, yet I make no apology
for I do consider this Dane to be a leading flyfisher and flytyer.

Jacobsen's dressing of the reed smut (*Simulium*) is very
effective and indicative of his careful dressings.

The hook, generally size 16, is fixed in the vice. Start the

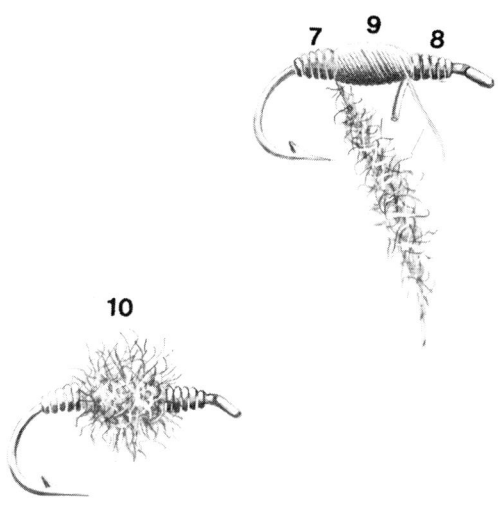

brown tying silk, 1, down the hook shank and immediately tie in a length of 0.15mm silver wire, 2. Continue the silk down the shank to the bend where a further length of the silver wire is tied in, 3. Return the silk in close turns back up the shank for approximately one third the length, 4, and tie in a length of copper wire, 5.

Wax the silk and dub on blood-red cow's hair, pre-treated with silicone floatant, 6. Wind the silver wire and form the rear third of the body, covering the tying silk, 7. Secure and cut off waste wire. Wind the forward silver wire back down the shank for approximately one third of its length, letting the waste end hang by the hackle pliers, 8. Now wind the copper wire backwards and forwards over the middle third creating a slightly tapered shape, 9. Trap the waste end of the forward silver wire under the copper wire, putting two half hitches in the latter to secure both. Remove waste ends of both silver and copper wire.

Wind the dubbed silk over the copper wire base, securing it just behind the forward silver windings, 10. Truly an excellent pattern.

PVC Nymph

Goddard

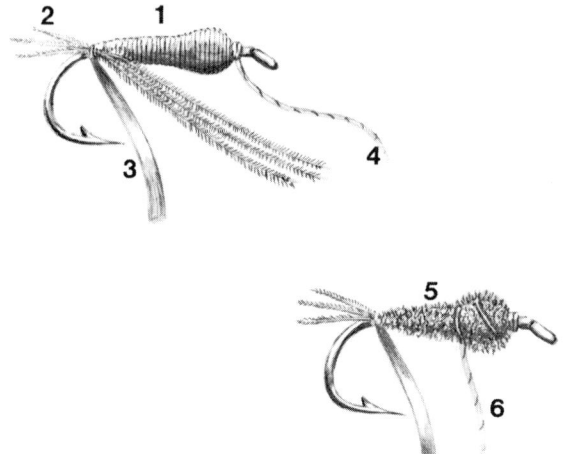

Each generation of flyfishers seems to produce just a few practitioners who have given to the craft far more than they have taken from it, G. E. M. Skues and F. M. Halford being just two examples. Men who have devoted hours of study to some particular facet of trouting and have then distilled their knowledge and observations between the covers of books for the benefit of those who have the desire to fish in print as well as in water.

High on the list of contemporary angling authors must be John Goddard, a skilled amateur entomologist, a superb photographer of insect life and no mean flydresser. The combination of these talents has produced two books that must be required reading for all who would know more about the trout's diet. I refer to *Trout Fly Recognition* of 1966 and his later work of 1969, *Trout Flies of Stillwater*.

Goddard's PVC Nymph is an excellent general representation of a number of the olive tribe and can be used with confidence on stream and still water.

The hook size varies between 12 and 16. Tie in a length of brown silk, followed immediately by a length of fine

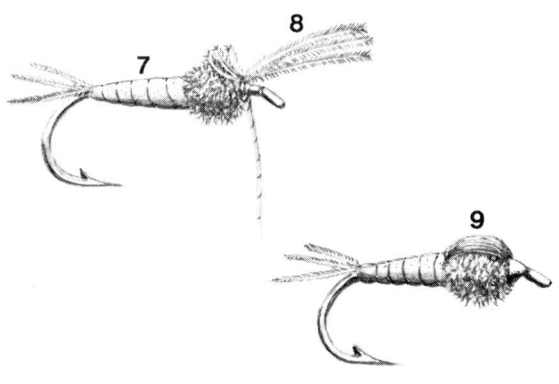

copper wire. Wind the wire backwards and forwards along the hook shank to produce a nymphal shape, 1. On your first trip with the wire to the bend tie in three golden-olive condor herls as the tails, 2, protruding no more than one-eighth of an inch. Now tie in a length of olive-dyed PVC, approximately one-sixteenth of an inch wide, 3.

Having formed the underbody shape secure the copper wire at the eye end with the tying silk, 4, and remove waste wire. Wind the condor herl over the copper wire to form the body and thorax, 5. Secure and remove waste ends. Take the silk in wide turns to the rear of the thorax, 6. Wind the PVC over the condor herls of the abdomen, 7, in slightly overlapping turns. Secure and remove waste material.

Take the silk back to the eye and tie in three blackish pheasant tail fibres, 8. These are then doubled and re-doubled over the top of the thorax to form the wing cases, 9.

To imitate the translucent effect of some nymphs one can wind fine silver wire in widely spaced turns between the herl and the PVC of the abdomen.

Pheasant Tail Nymph

Sawyer

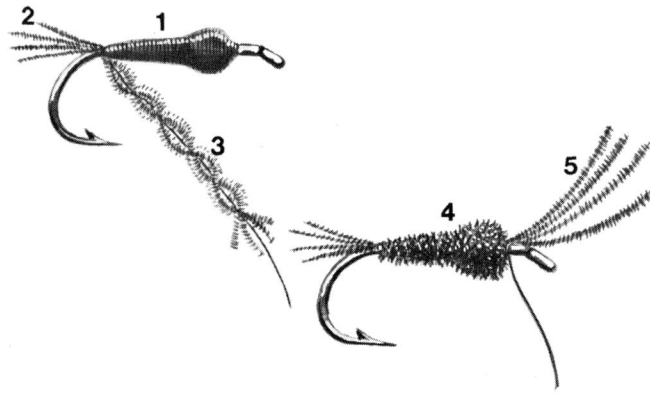

Few flyfishers who have an interest in nymph fishing can be ignorant of the name of Frank Sawyer and his long years of nymphal experimentation on the Wiltshire Avon. In truth the names of Sawyer and Avon are practically indivisible.

Born in 1906 in the village of Bulford he was, by 1926, the under-keeper to Frederick Martin, the head-keeper to Lt. Col. Bailey. Two years later he took over as keeper on the water controlled by the Officers' Fishing Association. In the same year, 1928, he became friendly with Brig. Gen. H. E. Carey, CMG, DSO, a most competent flydresser who put Sawyer on the road to the tying of fur and feather nymphs.

Some consider that Frank Sawyer and G. E. M. Skues were both engaged in the development of the artificial nymph on parallel lines. This is not so. Skues was particularly engaged with the emergent nymph, or the nymph just below the surface, while Sawyer was experimenting with the deep sunk nymph, hence his many references to weighted patterns. We do not have the space to investigate Sawyer's work in depth and so must content ourselves by describing what is possibly his most famous pattern. For

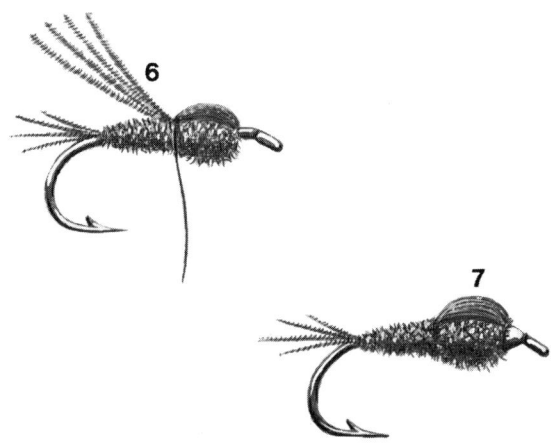

those who would know more about this highly interesting man and his work I recommend *Nymphs and the Trout* by Frank Sawyer, published in 1958 and revised in 1970.

The hook size is between 14 and 16. The pattern is unique in that no silk is used in the tying. Instead, fine red-coloured copper wire is used throughout. Wind the wire up and down the shank to form a nymphal shape, 1. At the bend tie in four fibres of deep reddish-brown cock pheasant tail feather, 2, the centre feather being the best. Twist the fibres and the copper wire together, 3, then wind the whole back up the hook shank and over the copper wire underbody, 4, allowing the waste ends to stand out over the eye on top of the shank, having carefully disengaged them from the wire, 5.

Take the wire in a wide turn to the rear of the thorax then bring the forward pointing herls back over the thorax to form the wing cases, 6. Take the wire forward and repeat the process as many times as required to form a pronounced case, 7. Snip off the waste ends of pheasant tail fibre and whip-finish with the wire.

You have now created one of the most effective nymphs in the angler's armoury.

Stonefly Nymph

Whitlock

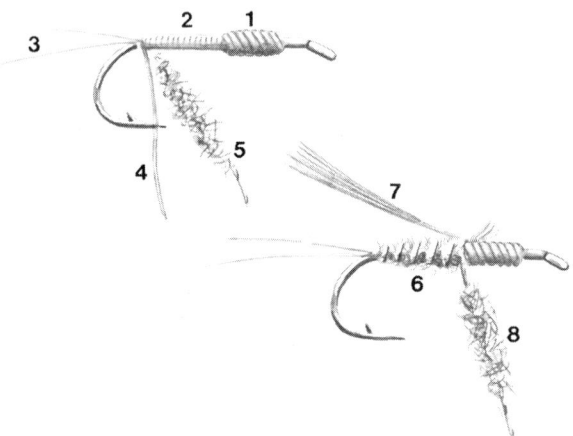

It has been my pleasure to enjoy the company of many of the top names in American flyfishing who have visited these shores and shared a rod with me. Generally they have all impressed me with their tremendous enthusiasm and river craft in all its aspects and, though I hardly dare to say it, their in-depth knowledge of entomology seems to be way ahead of that exhibited by the British angler. Few have impressed me so much, especially in the area of entomology coupled to imitative flydressing of a high order, as Dave Whitlock of Oklahoma, a professional flyfisher and lecturer in his own country.

While this pattern is based upon the American Stonefly I can assure you that it has survived its journey from the western waters of that great country to the streams and rivers of this pleasant land without too much surgery. No doubt if tied on the usual American hook size it would frighten the living daylights out of the British trout, but trimmed down to an acceptable size it is a most efficient pattern.

The hook size I have found most effective is between 10 and 14. Tie in a length of fine lead wire and wind onto the shank in the area of the thorax, 1. Continue the orange

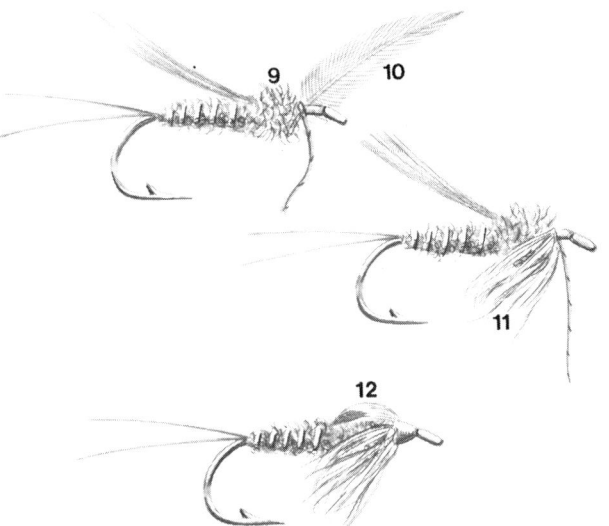

tying silk to the tail, 2, and at the bend dub onto the silk a very minute amount of rabbit fur. Wind this once round the shank then strip off the waste. Now tie in a doubled length of dark brown or black horse hair, 3. The small dubbing ball is to keep the two tails apart. Now tie in a length of fine gold wire, 4, and dub the tying silk with a mixture of quarter golden-brown rabbit or beaver fur, quarter dark brown seal's fur, quarter burnt-orange seal's fur and a quarter dark amber seal's fur, 5. Wind the dubbed silk to form the abdomen and rib with the fine gold wire, 6.

Tie in a strip of dark brown turkey secondary feather fibre, 7. Re-dub the tying silk, the mixture as before, 8, and wind in the pronounced thorax shape, 9. Also tie in a grizzle hackle dyed light-brown, 10.

Wind the hackle round the shank and by figure-of-eight turns of silk draw downwards and rearwards, 11. Bring the wing case material over the thorax and tie down, 12. Complete with a whip finish.

Whitlock specifies two antennae pointing over the eye but I doubt if these are of any value when hunting British trout.

Shrimper

Goddard

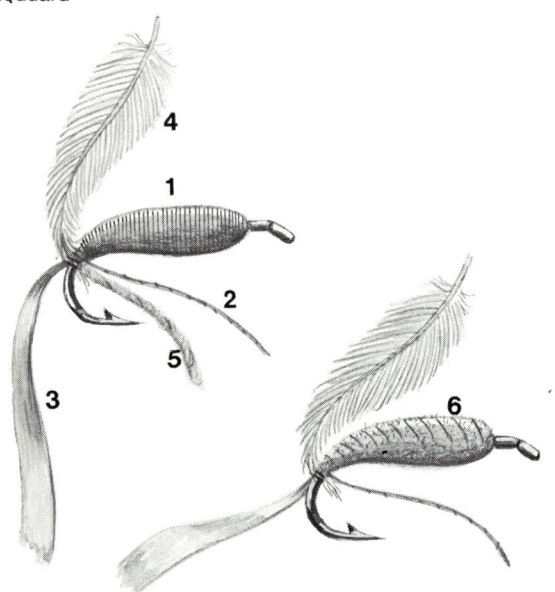

Another very good pattern by that top line angling ento-mologist John Goddard, this time a general representation of crustacea, a part of the trout's diet that at times is taken with such alacrity and helps in no uncertain manner in giving to the flesh of the trout that lovely pink hue. How unappetising some trout do look with their leaden grey or brownish flesh.

Developed with the stillwater flyfisher in mind this artificial should be fished on a floating or sink-tip line in the shallow margins near the bankside weed. The Shrimper should be retrieved by very slow movement of the line, with frequent pauses, as it lifts marginally from the bed of the lake.

The hook size is generally between 10 and 14. The tying silk of an orange colour is wound down the hook shank and well round the bend of the hook at which point a length of fine copper wire is tied in, this being wound back-wards and forwards along the shank to create a positively

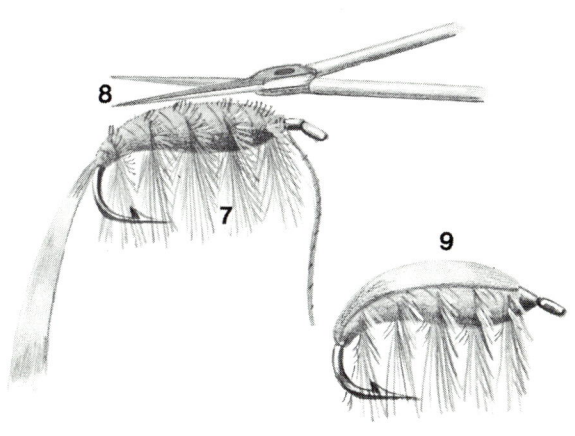

hump-backed shape, 1. On the final backwards pass the wire is secured by the tying silk, 2, and the waste end removed.

Now tie in on top of the hook a tapered length of PVC, approximately three-sixteenths of an inch wide, tapering at each end, 3. Follow this with a honey-coloured hackle, 4, and a length of marabou olive silk, 5.

Wind the marabou silk over the copper wire underbody, 6, and secure by allowing it to hang from hackle pliers. Take the tying silk in wide turns over the marabou silk and secure at the head. Remove waste marabou silk.

Wind the hackle in evenly-spaced open turns over the marabou silk body and secure, 7. Take a pair of sharp scissors and carefully cut away the hackle fibres from the top, and partially from the sides, of the body, 8.

Stretch the PVC tightly over the whole fly to create the back, 9. Complete the Shrimper with a carefully tapered whip-finish and then varnish the head.

Eric's Beetle

Horsfall Turner

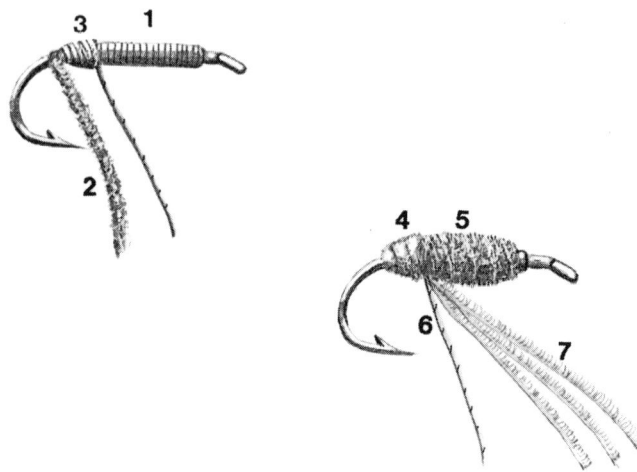

Whilst this book is mainly devoted to nymphal patterns I do not consider it wrong to include other related types and in my reckoning beetle patterns come within my own personal terms of reference.

This particular beetle, Eric's Beetle, was developed in 1940 by that most thoughtful flyfisher Eric Horsfall Turner, to my mind the most accomplished contemporary fisher of the fly for trout and salmon of flowing waters that it has been my pleasure to know and call a good friend.

The final beetle design was evolved on the river Derwent that flows through Hackness Valley, some short distance inland from Scarborough, where for many years Eric was the Town Clerk. This was no haphazard pattern developed by guess and by God. It was the result of much observation and experimentation. For example, prototypes had red wool tags but Horsfall Turner's carefully documented records confirmed without doubt that the yellow wool tag proved to be the most effective trout taker.

Eric's Beetle should be fished with the leader greased to within two or three inches of the leader point, allowing the pattern to sink that amount below the surface. Cast upstream, especially under bankside bushes, the fly can pro-

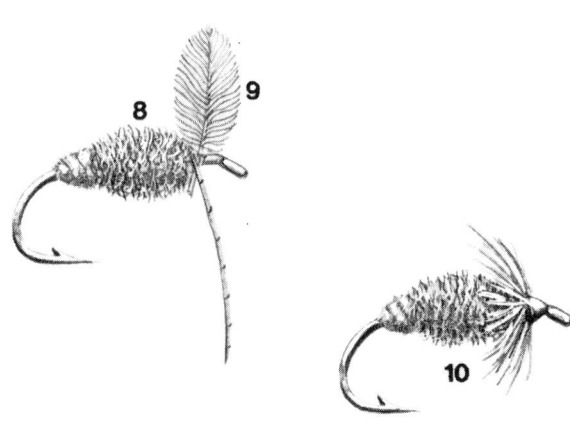

vide surprising results. Do not worry if it lands with a distinct 'plop' for its entrance into the water seems to have a good effect on the trout. My own fly box is never without a selection of this most effective pattern and I suspect that it will be a firm favourite for countless years to come.

The hook size is generally 10 or 12, though I have had success with this pattern when dressed down to size 14. The tying silk, 1, is either brown or black. Wind down to the bend and tie in a length of yellow wool, 2. Wind the silk in close even turns back up the shank for approximately one-quarter of its length, 3. Wind on the yellow wool to cover the silk, 4, and continue to wind the wool to form a plump underbody. Secure with backward and forward open turns of silk, 5, ending with the silk just in front of the yellow tag area, 6. Remove waste wool.

Now trap in a bunch of peacock herls, 7, and wind them backwards and forwards over the yellow underbody, making sure you do not encroach onto the tag area. Take the tying silk forward in wide open turns and secure the herl, removing the waste ends, 8. Tie in a very small feather from the base of a starling's wing, 9, and wind onto the shank, two turns at the most. Secure, remove waste feather, and complete with a whip finish, 10.

Iron Blue Nymph

Skues

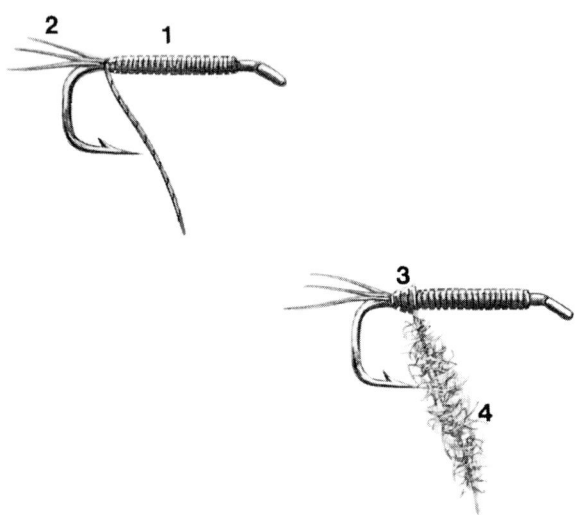

Possibly the most famous nymphal pattern of all time developed by that *Meisterfischer* of subsurface feeding trout, G. E. M. Skues. I doubt if there is a flyfisher in the world who has not heard of the old gentleman. If there is such a person then I venture to suggest that he does not qualify for the title of flyfisher for his education is most sadly lacking.

Skues can rightly be called the father of nymph fishing. Certainly the craft of fishing the upstream wet fly had been indulged in countless years before Skues strode the banks of the southern chalk streams but he was the first to systematize nymph fishing and develop the artificial as a true representation of the natural nymph. He was also a steadying influence in a turn-of-the-century world that had gone mad on F. M. Halford's doctrinal teachings on the dry fly. The nymph versus dry fly war reached its zenith with a most serious debate of these two seemingly opposed methods at the Flyfishers' Club shortly before the Second World War. Happily the controversial days are behind us, though pockets of resistance still fight on in the depths of Hampshire!

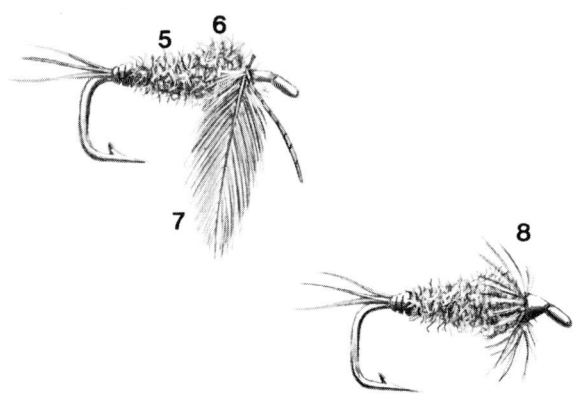

The fly described here must be Skues' best known pattern. You will also note that it is the only fly I have illustrated tied to a sneck-bend hook. Why? Well, the old master dictated so and such is my respect for him that I could not bring myself to draw any other.

Skues specified a number 16 sneck-bend hook and so it shall be. The red tying silk waxed with brown wax, 1, is taken down the shank to the bend where three soft white hackle fibres are tied in as tail whisks, 2. Wind the silk back up the hook in close even turns, 3, and then dub the silk with a thin layer of mole's fur, 4.

Wind the dubbed silk carefully up the hook to form a gently tapering abdomen, 5, and a thickening thorax, 6. Now tie in a very short-fibred nearly black hackle from the throat of a jackdaw, 7. Wind this feather hard up against the thorax, 8, no more than two turns maximum. Complete the nymph with a carefully tapered whip finish, varnished.

Go to the stream and cast to 'that little brown wink under water', sure in the knowledge that you are using a most effective, and historic, pattern.

Bow Tie Buzzer

Sawyer

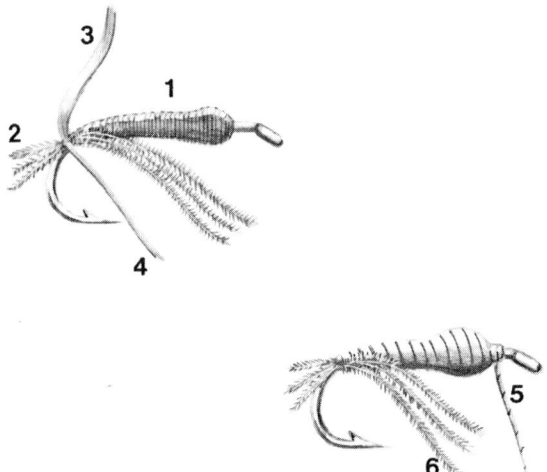

Another pattern devised by that thoughtful flyfisher of the Wiltshire Avon, Frank Sawyer. His knowledge of underwater fly life is quite encyclopaedic and one can be sure that the patterns he has perfected have a long pedigree of careful testing. While his name is for ever linked with the streams of Wiltshire he has not neglected the stillwater aspect of flyfishing. This is proved by the Bow Tie Buzzer, for it is a cracking imitation of the hatching larvae, with a most unique way of imitating the white celia.

All other patterns that I know imitate this aspect of the natural by tying in some appropriate material at the head of the fly. Sawyer chooses the alternative of making the celia part of the leader (cast), believing that the pattern when hanging in the water in the correct manner, almost vertical—caused by the way the leader point is threaded through the eye—can be made to spin by gentle retrieve action on the line. A most interesting conclusion, not arrived at in five minutes.

The fly is generally tied on a size 12 hook. Again we discover Sawyer's use of fine copper wire in place of tying silk, this time of a golden colour. Wind the wire up and down the hook shank until you have created the correct

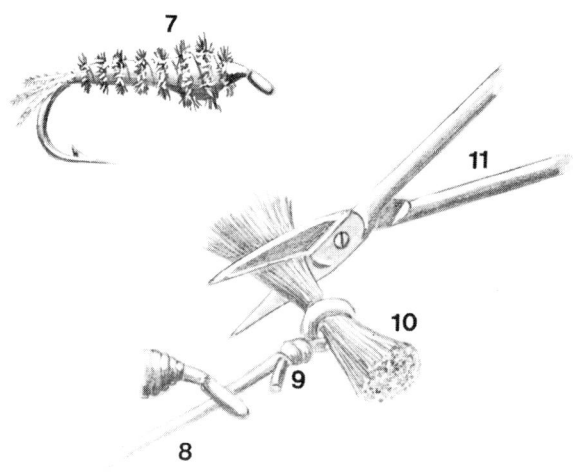

shape, 1, ending at the bend. Tie in three cock pheasant tail feather fibres, 2, but do not cut off the waste ends, they are required later, in the making of the fly. Now tie in a length of flat silver tinsel, 3. Take the wire, 4, back up the body to the eye, followed by the flat silver tinsel in even, closely-fitted turns, 5. Tie in and cut off the waste ends. Twist the pheasant tail fibres together, 6, and wind in open turns up the body of the fly so that the flat silver tinsel shows through, 7. It is in fact an advantage to wind the wire and herls together for strength.

We now come to the innovative aspect of this pattern, the celia. Pass the point of the leader through the eye of the hook *from the underside*, 8, this will ensure that the fly hangs almost vertically in the water. Via a loop knot, 9, tie in a bunch of white nylon wool fibres, 10. Pull the knot tight and cut off the excess wool on either side of the knot with a pair of sharp scissors, 11.

I must confess that I have not the patience to fish this pattern with any degree of regularity, but Sawyer has never produced a pattern that would not take trout and I suggest our confirmed stillwater addicts should give it a try.

Marabou Bloodworm

Price

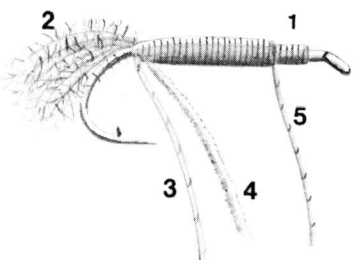

An excellent pattern of the bloodworm larvae developed some five years ago, after considerable experimentation, by Taff Price. He considers that the majority of patterns imitating those creatures of the genus *Chironomus* lack the integral movement so evident in the natural. Price's method of simulating such movement is simple but original.

It is important that the artificial be allowed to imitate the natural species by fishing it in the correct manner. It is best fished near the bottom with a medium speed recovery, alternating with pauses and an uplifting of the rod. These actions cause the delicate marabou tail component to wiggle in a most lifelike manner.

The pattern is tied on long shank hooks between 10 and 14. Take the tying silk of a red colour, 1, down the hook shank in close even turns to the bend of the hook and tie in a bunch of red marabou feather fibres, 2, followed by a length of fluorescent red silk for the ribbing, 3, and a length of red floss silk, 4.

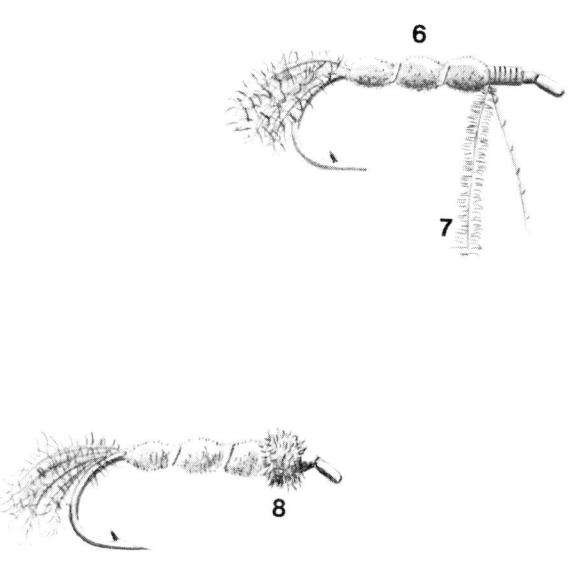

Return the tying silk in close even turns back up the hook shank to a point as indicated at 5. Wind the red floss body material up the body, aiming to create an undulating effect, and tie in. Rib the body with the fluorescent red silk and secure as at 6. Cut off waste ends.

Tie in a length of peacock herl, 7, and wind in the space between the body and the starting point of the tying silk. Secure and cut off the waste end, 8. Whip finish and carefully varnish to complete the fly.

Price is most honest when he states that the peacock herl serves no great purpose but gives the pattern an air of respectability! I suspect that he is still suffering from the effects of Halfordiana!

A green version of this pattern is also effective, as may be a brown variant. The naturals have also been known to be virtually colourless; not easy to imitate.

The Chomper

Walker

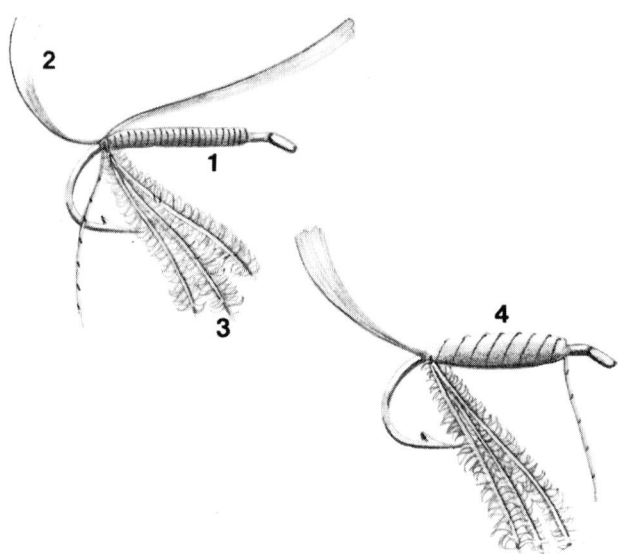

Few men are as active in the area of angling journalism as Richard Walker and to the vast majority who take their pleasures on the river bank (fishing pleasures!) Walker is a household name. His regular articles on all aspects of fishing, and his books, are always of interest and very frequently controversial, but I have long held the view that Richard Walker loves a good argument in print from which everyone can draw some information and enlightenment, He certainly knows a thing or two about the craft of flytying and many a newcomer to flydressing must have been helped by his words.

For the beginner his Chomper series of patterns have considerable merit, for they not only take trout but are also quite easy to dress. Nothing gives the new fellow more confidence than to take a fish on a fly of his own tying.

The colours of the Chomper family are only restricted by one's daring, but I will list some of the more popular colour combinations after the tying instructions.

The hook size is what you care to make it, though 14 to

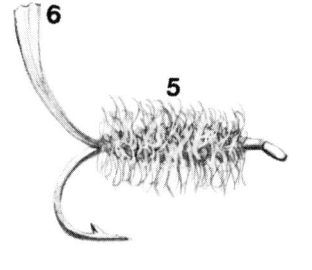

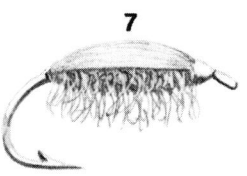

10 are popular sizes. Tie in the silk, 1, and wind in even turns down the hook shank to the bend. Tie in a length of Raffene, 2, followed by the ostrich herl body material, 3. Take the silk backwards and forwards along the shank to form a plump tapered underbody, 4.

Twist together the herls and wind up the shank to form the body, 5. Damp and stretch the Raffene that is lying out from the bend, 6, and pull it over the top of the body to form a shell, 7. Secure with the silk and cut off the waste. Complete with a well varnished whip finish. Do resist the temptation to varnish the Raffene back, for its translucency is part of the fly's attraction.

Try your hand at these:

Sedge Pupa   Dyed amber or green herl with a light brown Raffene back.

Corixa   White herl with light brown or green Raffene.

Beetle   Black herl with black Raffene.

Bare Hook Nymph

Kite

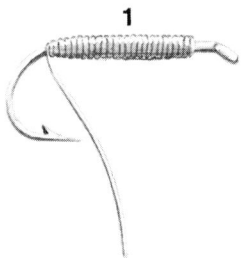

1

Flyfishing occasionally produces a personality who becomes extremely well known in a comparatively short space of time. Major Oliver Kite (1920-1968) was such a person and he developed a considerable following via his highly successful TV series, his many articles to such magazines as *Trout and Salmon* and his regular weekly column in *Shooting Times*.

All who write have their critics and 'Ollie' was no exception, but I do firmly believe that he was instrumental in causing countless anglers to take a more observant and intelligent interest in their river craft, thereby becoming far more effective takers of trout.

His most famous pattern was the Kite's Imperial, a dry fly that I suspect I use more than any other pattern when fishing the floater and a fly that will be featured in a further book in this series. Here we are concerned with nymphs and Kite's Bare Hook Nymph must be the simplest pattern ever devised and, if one has the courage to try it, a most effective nymph when fished in the correct manner.

The idea for the B. H. Nymph came to him when fishing a very badly chewed Pheasant Tail. Most of the herl had long since gone, leaving only the copper wire underbody, and yet the trout would take it with enthusiasm. Of course

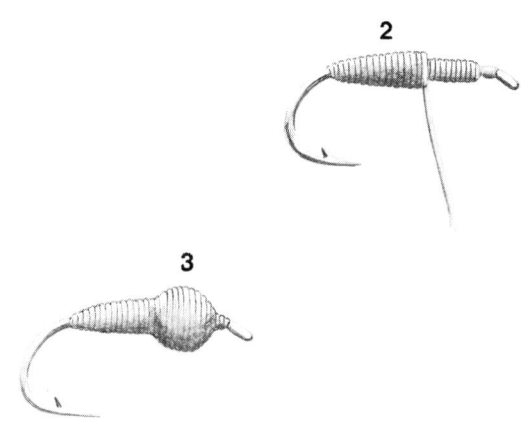

the method of fishing the fly enters into the picture, and Kite believed that if the pattern was presented some feet above the trout and then when it was almost in front of the fish it was lifted to simulate a rising nymph, the trout would take it. This technique he called the 'induced take'. I hasten to add that he did not claim originality in the technique.

I came to know Oliver Kite in his latter years, and as I type these words I can look up and see framed examples of his patterns, including the Bare Hook Nymph, that he tied for my collection.

To tie this nymph place the hook, either 14 or 16, in the vice and wind a length of copper fuse wire down the shank to the bend, 1. Return the wire up and down the shank to create a tapered abdomen, 2, followed by a considerable build-up for the thorax, 3. Finish off and there you have the most austere fly you are likely to tie. (Kite also tied this pattern with just a thorax of copper wire.) Remember, it is the way it is fished that gets results.

Olivér Kite died on the afternoon of 15 June 1968 while fishing the river Test, and hoping to repeat an enjoyable morning's sport. Can there be a better way to go? I think not.

Gold Ribbed Hare's Ear

Originator unknown

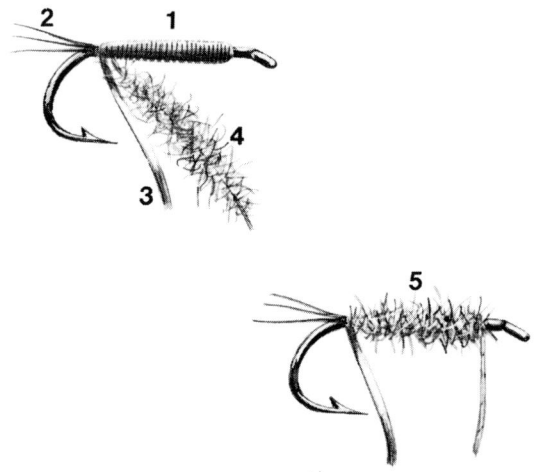

This must be one of the best known patterns in the fly-fisher's collection and yet it is most difficult to try and pin down, with any degree of certainty, when it was first used. One suspects that such a simple dressing must have been around for many many years, though never recorded until years after its birth.

The Gold Ribbed Hare's Ear caused a great deal of soul-searching on the part of that celebrated dry fly man F. M. Halford (1844-1914). His commonsense told him that it was a first class pattern, while his self-imposed code of 'dry-fly-only' would not allow him the pleasure of its use unless it was bastardized into dry fly form by the addition of starling fibre wings tied upright. It will of course take trout as a floater but how much better it is when used in its true nymph form.

Flyfishers are generally agreed that this pattern seems to be most effective when fished in the surface film as a nymph at the point of hatching. We can narrow down its effectiveness still further by saying it is the ideal pattern

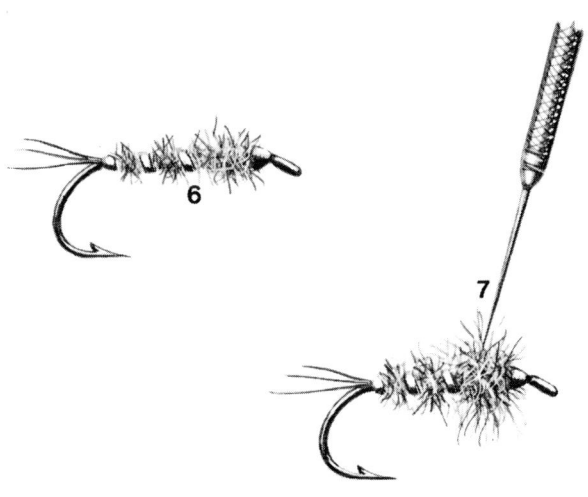

when a number of the olives are hatching.

To keep the nymph in the desired position in the water grease the leader to within an inch or so of the point. I know it will take trout when fished downstream, but it is far more effective when cast upstream to nymphing trout.

The dressing I now describe is, I venture to suggest, the nearest to the original pattern.

The hook size is between 14 and 16. The primrose tying silk, 1, is taken down the hook shank in close even turns to the bend when three short dark fibres of hare's ear fur are tied in for the tails, 2. Follow the tails with a length of gold flat tinsel, 3. Now rewax the silk and thinly dub with the darkish fur from the root of the hare's ear, 4.

Wind the dubbed silk up the shank, 5, and follow with open turns of the gold tinsel, 6. Put on an extra turn or two in the area of the thorax. Whip finish and varnish the head. Now take a dubbing needle and gently tease out some of the longer fibres of hare's fur, 7.

August Dun Nymph

Clegg

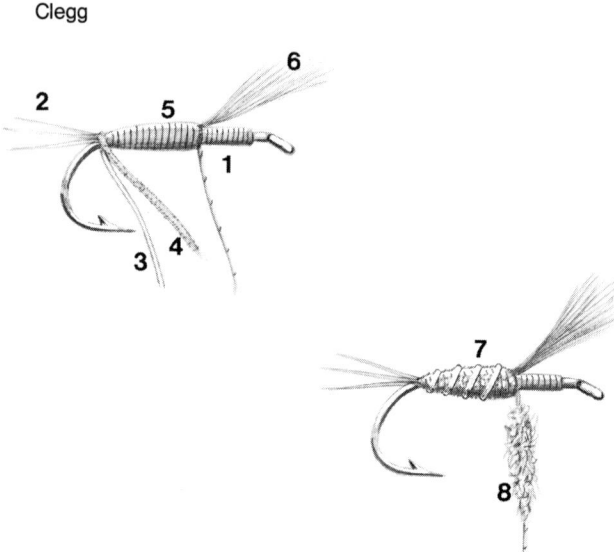

One of the most thoughtful and inventive flytyers in the United Kingdom must be Thomas Clegg of Scotland. His books *The Truth About Fluorescents*, *Hair and Fur in Fly Dressing* and *Modern Tube Fly Making* are required reading for those who would be craftsman flydressers and he has opened the eyes of many to the restrained use of fluorescent silks. No less an authority than Eric Horsfall Turner considers Clegg to be well qualified to be our top flytyer.

I have used patterns devised by Thomas Clegg over a number of years and have finally overcome my natural reactionary attitude to the use of fluorescent materials, now believing that their use, *in moderation*, can enhance the fly in the eye of the trout.

Clegg's tying of the August Dun Nymph is not difficult, but do take care to form a shapely replica. The hook size is between 12 and 14. Take the light green tying silk, 1, down the hook shank to the bend, at which point tie in three short strands of guinea-fowl fibres dyed yellow, 2. Now tie in a length of copper wire for the ribbing material, 3, and a

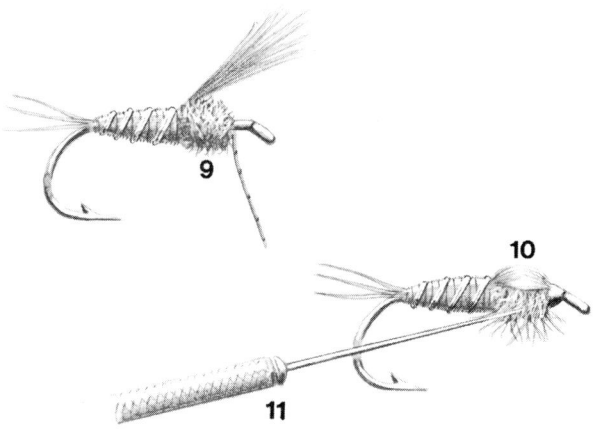

length of signal-green 'Depth Ray Fire' fluorescent floss, 4.

Wind the silk up the shank to abdomen length, 5, tying in a bunch of red squirrel tail hairs that will form the wing cases, 6.

Take the signal-green DRF and wind carefully to create a tapered abdomen, and rib it with the copper wire, 7. Re-wax the tying silk and dub with grass-monkey fur of a blue-dun/yellow-olive hue, 8. Wind the silk and dubbing to create the thorax shape, 9. Pull the wing case hairs over the thorax, doubling and re-doubling if required, to form the cases, 10. Complete the head with a whip finish and varnish. Now take your dubbing needle and carefully pick out the longest of the grass-monkey fur for the legs.

Experiment with fluorescent materials by all means, but do not think that the more you use the better the fly. A pattern built entirely of DRF would stand out like a Las Vegas hooker at the vicar's tea-party. All you need is a faint hint of these bright substances.

Footballer

Bucknall

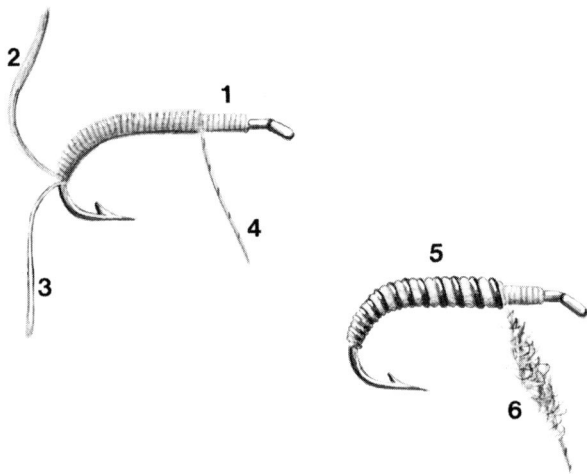

Geoffrey Bucknall is very well known to the flyfishing world as a knowledgeable angler, a well-read author of numerous books on flyfishing and a highly skilled tyer and inventor of very good fly patterns. (One can also add a successful tackle-dealer and businessman.) However, I think that his name will always be most closely linked with stillwater flyfishing and one pattern of fly in particular, his famous Footballer.

The nymphal pattern was not the result of some whim, a fly tied on the spur of the moment that by mere chance happened to be very attractive to the trout. No, it was the end result of countless hours examining the stomach content of many fish and the careful study of blown-up photographs of the buzzer pupa.

Purists . . . aye, they do exist on still water, chalk streams not having a monopoly . . . may consider that the pattern is incomplete because the filaments over the eye are absent. This omission is deliberate for Bucknall believes that nothing must impede the swift entry of the nymph into the water when casting to a cruising trout.

Since the first Footballer, described here, there have

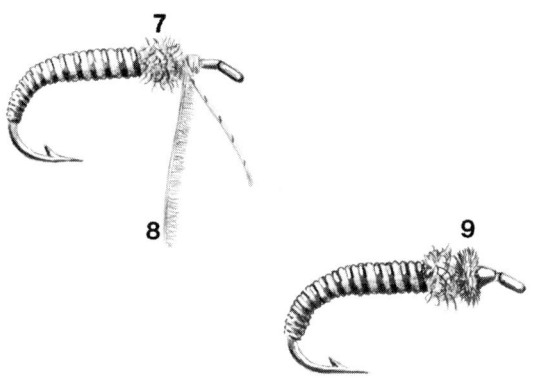

been many variations on the original colour theme; red, olive, black, white, orange, green, etc. Fluorescents have also been tried. Despite the variations many stillwater fly-fishers swear by the old original pattern and it does account for a most impressive number of trout taken from venues up and down the country.

The hook size is generally between 8 and 14. Take the tying silk, 1, my usual choice is white for the original dressing, down the shank to a point well round the bend. Tie in a length of black horse hair, 2, and a length of white horse hair, 3. Return the tying silk in close even turns up the shank to the point shown, 4.

Very carefully wind the black and white horse hair in alternate bands, each hard up against the other, 5. Tie off with the silk and remove the waste ends.

Re-wax the tying silk and dub with mole's fur, 6. Wind on in a forward direction, remembering to leave plenty of room at the eye, 7. Now tie in a length of peacock herl, 8, and wind on as a small head, 9. Complete the fly with a well-varnished whip-finish.

Orange Nymph

Price

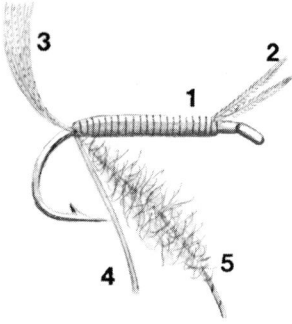

Another excellent pattern from the Taff Price stable. Created in 1973 its devising was prompted by a most frustrating stillwater afternoon when frantically feeding trout would look at nothing that Price had in his fly box.

In desperation he tied onto his leader point that grand old Yorkshire standby of countless anglers down the ages, the Partridge and Orange. It got him a fish. A friend of Price's, also in a desperate situation, finally took a trout on a Grenadier nymph, the interesting thing being that both patterns have orange bodies. A stomach content examination of the two victims revealed a glutinous mass of pale strawberry-jam coloured daphnia. The Orange Nymph was born from that examination and has since proved highly effective.

The hook size is generally 14, though it is equally effective down to such a size as 27, if you dare to fish so fine. Attracting is one thing, hooking firmly on such small hooks is another matter.

Start the orange tying silk down the shank of the hook and after a couple of turns, 1, tie in two or three fibres of unclipped swan feather fibres dyed deep orange, 2. Con-

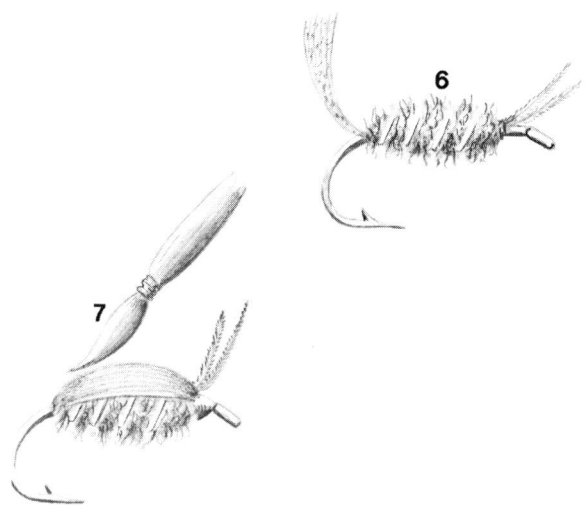

tinue the silk in close even turns to the bend, trapping the waste ends of the swan fibres en route.

Tie in a further bunch of swan fibres dyed deep orange, 3, followed by a length of gold wire for the ribbing, 4. This can be oval if preferred, or in fact can be omitted, for the lack of a rib does not seem to detract from the effectiveness of this nymph. On balance, and purely from the tyer's point of view, I would include it.

Re-wax the tying silk and dub with orange seal's fur, 5, or if you find difficulty in handling this rather springy substance use polypropylene fibres of the same colour. Wind the dubbed silk up the shank to form a body tapered at both ends, then rib with the gold wire, 6, securing it with the silk and removing the waste end. Remove any surplus dubbing from the silk.

Bring the swan fibres from the tail end and pull over the top of the body. Secure with the silk, being careful not to trap the forward pointing fibres, and complete with a tapered whip-finish. Carefully apply a coat or two of clear varnish to the swan fibres over the back, 7.

Leaded Chomper

Walker

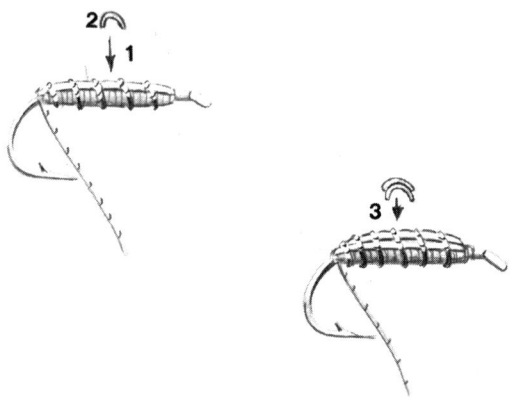

We have already examined the Chomper series developed by Richard Walker and now we come to the leaded variant of that pattern. The object of the lead understrip is to make the pattern sink quickly, far more quickly than it would with an underbody of copper wire, especially when a strong cross-wind is acting upon a floating fly-line causing such minimum weight nymphs to be dragged near the surface.

Another feature of this style of dressing is that the layers of lead strip on top of the hook shank cause the pattern to work bend and point upwards, diminishing the chance of fouling weed, moss, stones, etc. on the bed of the reservoir.

The method of tying is somewhat finnicky but well worth the time and effort. I have not detailed the colours of the materials for Chompers can be almost any colour that takes one's fancy, and they still take trout! The illustration is intended simply to show the tying sequence.

The hook sizes are as for the standard Chomper. Take

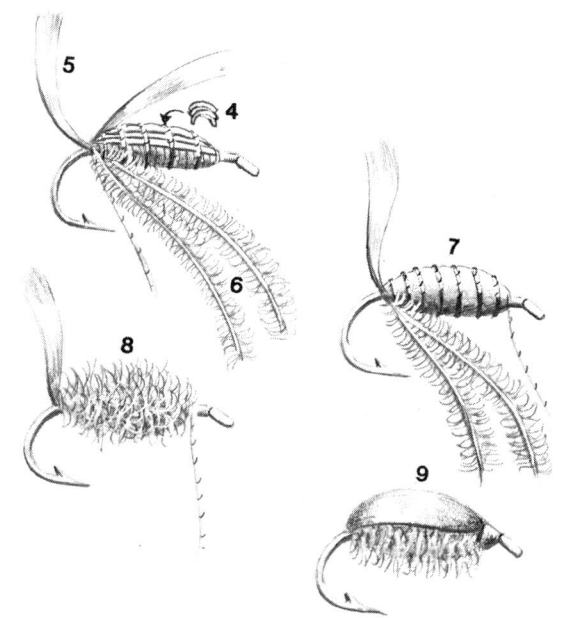

the tying silk down the hook, 1, and tie in on top a narrow
strip of lead foil from a wine bottle cap, long enough to
cover from the bend to the eye and wide enough to wrap
round half the diameter of hook shank, 2. Bind down with
the tying silk. Repeat with another strip of lead foil that is
a little shorter and a little wider than the first, 3, again
binding in place with the tying silk. The last strip is as long
as the first but has each end trimmed to a point to help with
body taper, 4. Bind down.

At the bend tie in a length of Raffene, 5, and the body
herls, 6. Wind the waste end of the Raffene over the lead
strips and bind down with the silk, 7. Remove the waste.
Take the herls in close turns over the Raffene and tie in, 8,
removing the waste ends. Damp and stretch the remaining
Raffene over the back of the hook, completing the fly with
a whip finish after removing the surplus Raffene.

As with the standard Chomper, resist the temptation to
varnish the back of the Raffene.

Tup's Nymph

Leisenring

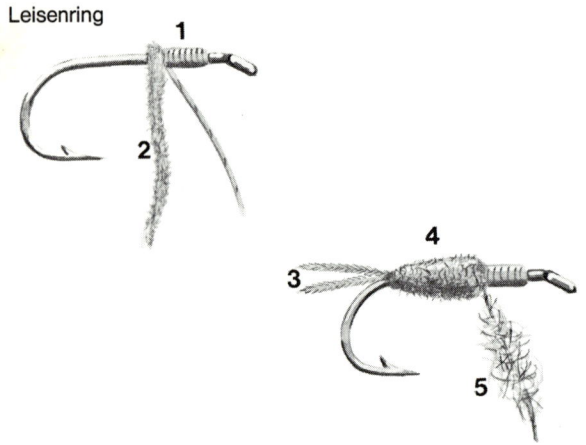

At the risk of being lynched I am prepared to go on record as stating that the American flyfisher has a greater interest in the life cycle of the fly as applied to trout fishing than his British counterpart. They may well have started later than us in fishing for sport but they have more than made up for the lack of years by intensive investigation into fly design and trout reaction.

One man who has become known as the American Skues was James Leisenring (1878-1951), a Pennsylvanian of German origin who fished the famed Brodheads and the limestone streams of his native county, waters that are the nearest thing in that vast country to our chalk streams.

He knew well the dry fly cult that swept America, as it also raged through England, headed by Halford and his disciples at the turn of the century. Leisenring was too knowledgeable an angler to go with this mainstream and continued to believe, quite correctly, that the trout took the major part of its food under the surface.

For a number of years Leisenring was in correspondence with G. E. M. Skues, discussing nymphal behaviour, and these two anglers had a lot in common. A copy of Leisenring's book on wet fly fishing, dedicated to Skues and signed by the American, is in the library of the Flyfishers' Club to this day.

Leisenring developed a number of nymphal patterns

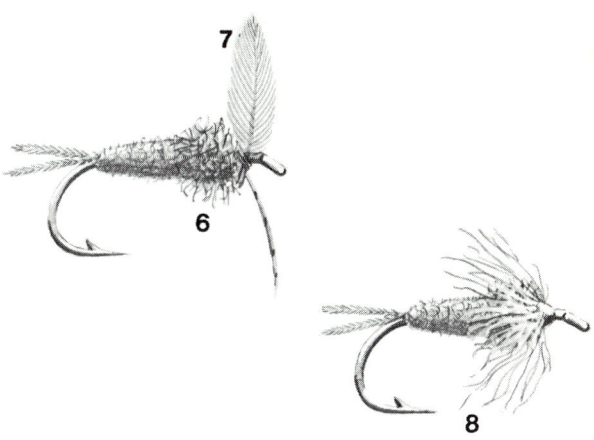

based upon his dictum that, 'You must tie your fly, and fish your fly, so that the trout can enjoy and appreciate it.' He also seems to have appreciated the technique of the 'induced take' long before Oliver Kite came up with the term.

This pattern is typical of his dressings. The hook is either size 14 or 15. Start the claret tying silk, 1, down the hook shank for the approximate distance of the thorax. At that point tie in a length of primrose marabou silk, 2. Wind the latter down the bare hook shank and at the bend trap in two cock pheasant tail fibres, 3. These are not essential really to the dressing for the originator usually omitted them. Wind the marabou silk back up the shank to form a tapered abdomen, 4. Cut off the waste end. Re-wax the silk and dub with well mixed claret and yellow seal's fur, 5. Wind this on as a thorax, 6. Now tie in a light-blue or medium dark honey-dun hen hackle, 7, winding it so that the fibres lie well back over the thorax, 8. Using the 'induced take' method these hackle fibres move in a most attractive manner, as the nymph is lifted in the water just in front of the trout. Complete the fly with a well-varnished whip-finish.

As an experiment try painting the bare hook shank under the marabou silk with white paint.

Black Nymph

Bridgett

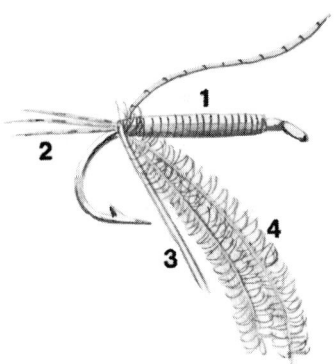

For many years the pages of the *Trout and Salmon* magazine were enlivened month by month with a regular series of fly patterns, written by the late Tom Stewart. In his series he described patterns old and new, ranging from tiny nymphs to large salmon flies. I doubt if a single reader of that magazine had the experience of dressing and fishing with all the patterns Stewart described, though many fly-dressers found something in the series that appealed, was tied, tried and became a firm favourite.

Such was the case for me when I first read about the Black Nymph. Stewart told us that the pattern was developed by the late Mr. R. C. Bridgett, a flyfisher who in Stewart's opinion knew more than most about the theory and practice of loch fishing. Bridgett recommended that the pattern be used as the tail or point fly on a wet cast and that it be dressed with long spider type hackles of the traditional North country wet fly.

I, in a heretical frame of mind, dressed my first Black Nymph with a very short-fibred hackle, typical of a Skues nymph, and never regretted meddling with the original de-

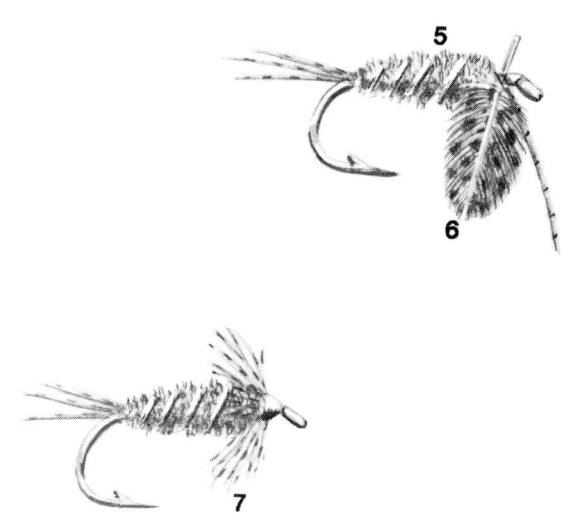

sign, for the fly has consistently taken trout for me from the Scottish lochs and burns to the southern chalk streams.

The original pattern was tied on hooks between 14 and 12, though with my own version I do prefer 14 to 16. Wind the black tying silk, 1, down the hook shank to the start of the bend. Tie in three short fibres from a well marked guinea-fowl feather for the tail, 2, followed by a length of fine oval silver ribbing tinsel, 3. Now tie in a strand or two of black ostrich herl, 4.

Wind the herl up the hook shank in a gently tapering form, followed by wide turns of the ribbing material, 5. Cut off the waste ends. If you wish to tie the original pattern now tie in a long-fibred guinea-fowl feather, well speckled. If you are interested in my variation tie in a very short-fibred guinea-fowl feather, 6. Wind the hackle, two turns at the most, 7, and complete with a tapered and well-varnished whip-finish.

I would not even hazard a guess what the pattern, original or my variant, might represent. All I do know is that it is a first class general pattern.

Red or Green Larvae

Goddard

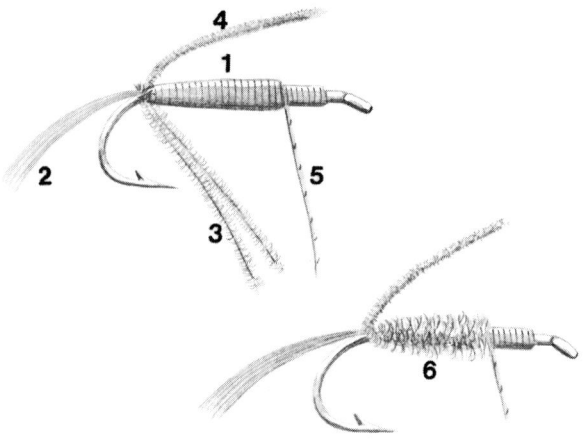

Yet another excellent pattern from the flytying vice of John Goddard, possibly the top amateur flyfishing entomologist of the present day. His accurate observation, coupled with his ability to translate fur and feather into working patterns, make him a man whose statements should not be ignored.

This simple dressing . . . oh, how often the most effective are the most simple to construct . . . was devised to simulate the larvae of the large chironomids, whose larvae are usually found in quite large numbers in the mud and silt of lakes and reservoirs.

The natural action of these larvae is quite difficult to imitate—a lashing figure-of-eight movement—though they also lie almost inert on the bottom.

Goddard uses a most interesting method of animating the pattern by the use of a curly piece of red ibis quill fibre which straightens out and re-curls as the artificial is twitched along. The red or green larvae should be fished on a sinking line, allowing the fly to lie upon the bottom of the lake, disturbing it with the occasional twitch of the line.

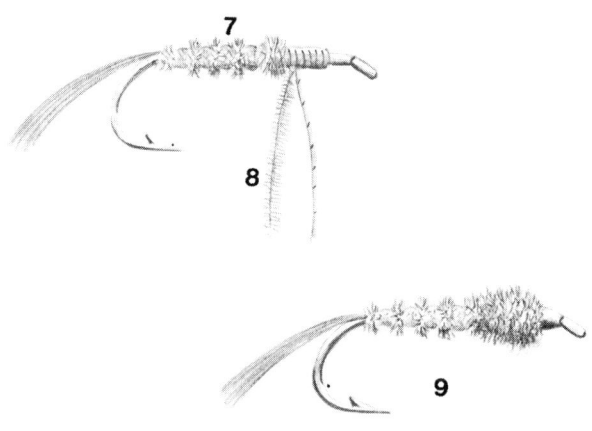

22

The hook size is between numbers 8 and 12, long shank. The brown tying silk, 1, is wound down to the bend at which point a piece from the curly section of the red ibis quill is tied in, 2. Alternatively you can use a red or green cock hackle with a very pronounced curve to the feather, or even two or three dyed fibres of heron or condor. The ideal is of course the red ibis.

Follow the ibis with two or three crimson or olive dyed condor herls, 3 – suitably dyed turkey is an acceptable substitute for condor herl – and a length of fluorescent floss silk of the same colour, 4.

Wind the brown tying silk back up the hook shank for approximately the length shown, 5, followed by the condor herls, 6. Now wind the floss in open turns, allowing the condor herls to spring through, 7. Cut off the spare end of the floss and the herls.

Now tie in a length of buff coloured condor herl, 8, and wind it to form an elongated thorax, 9. Finish off the fly with a varnished whip-finish.

Palmer Nymph

Voss Bark

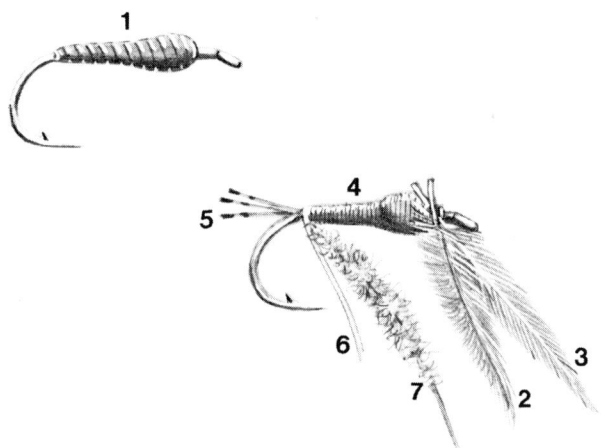

Conrad Voss Bark, one time political correspondent for BBC TV, is a well known figure in the higher reaches of flyfishing. His occasional articles in the angling press are always of great interest, concise, well written and thought provoking. His interests cover all aspects of flyfishing from the moorland rivulets of his West country home to the majestic streams of the chalk country, not forgetting the stillwater locations throughout the land.

Voss Bark's thoughts on the latter appeared in his well-received book *Fly Fishing for Lake Trout*, published in 1972, and one artificial to feature in that book is the Palmer Nymph, now a favourite of mine.

Palmered artificials are not new, and I hasten to say that Voss Bark never claimed exclusivity, for the flytying literature of the past is littered with palmer dressed wet flies. Skues put it rather well when he ventured the opinion that nymphs so dressed gave a 'buzz effect', i.e., movement to represent the activity of the gills and legs. Contemporary fly designers seem to be ignoring this lifelike sparkle and more thought should be given in this direction.

Voss Bark's pattern is devised to be fished either in the subsurface film or quite deep. It should be fished quite

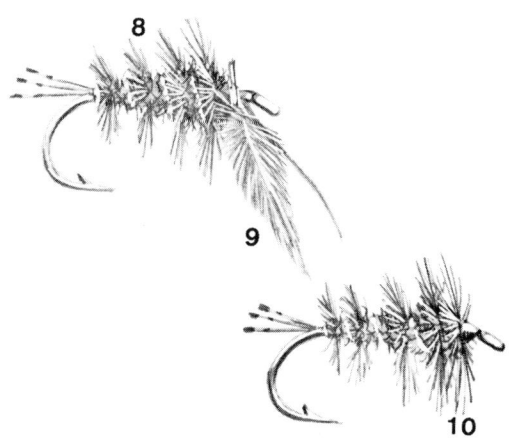

slowly with an occasional burst of energy imparted by a twitch of the line to activate the body hackle.

The hook size is 12, or for deep fishing up to size 8. Firstly wind on the shank lead or copper wire in the nymphal form, 1. Follow this with primrose tying silk, covering the wire underbody and ending up back at the eye. Tie in two short-fibred cock hackles, one golden-olive, 2, and one red, 3. Return the silk down the body, 4, and at the bend tie in a short length of golden pheasant tippet, or topping, for the tail, 5, followed by ribbing material of fine gold or silver wire, 6.

Dub the tying silk with a mixture of yellow and green seal's fur, arranged so that it darkens towards the thorax, 7. Wind the dubbed silk up the body and secure. Wind the two hackles down to the bend. Keep tension on the hackles and wind the ribbing through them to the eye, 8, trapping them en route. Twitch off the waste ends, after they are secured by the silk, and remove the waste ribbing.

Tie in a reddish coloured head hackle, 9, and wind two or three turns in front of all, 10. Complete with a well-varnished whip-finish.

Gerroff

Goddard

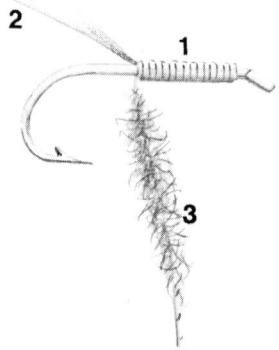

What odd names are given to some artificial patterns and yet, usually, there is some reason behind the christening. This nymph was designed by John Goddard and named by that other great flyfisher and author Brian Clarke. Why 'Gerroff'? You will find out shortly.

The nymph was designed originally for river fishing, especially to combat the drought conditions of 1976 when the water flow was reduced to a bare minimum on most streams. The usual weighted patterns sank too quickly in near-stillwater conditions while the unweighted nymphs that tended to stay in the surface film did not attract the trout, especially of the river Kennet, when they stayed glued to the bottom.

Goddard decided on a shrimp pattern and to achieve a slow sink rate dressed it on only half the shank length. He admits that he does not know if it is because of the materials used or the pattern/hook size relationship but the fly has proved to be one of the most successful ever devised by that talented angling entomologist. In fact, on its first baptism of fire it took twelve trout weighing 49lb to two rods and since that day has proved its worth on stillwater and stream in no uncertain manner. The inventor states that he does not recommend the pattern for very large, deep, stillwater locations, but on small, clear still waters it is really killing.

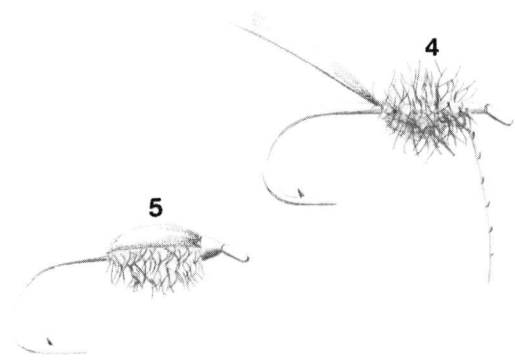

24

It should be fished on the point of the leader, on a floating line, and cast to a feeding trout or alongside weed beds. Goddard considers that the distinctive 'plop' of the entry frequently brings trout from many yards distant. He has also observed a fish take the artificial, then reject it, three or four times, proving it to be attractive.

The tying is simplicity itself, ideal for the tyro. The hook size is between 10 and 14, slightly longer in the shank than a standard pattern. Wind the brown tying silk, 1, half way down the hook shank. Now tie in a length of clear PVC strip, 2. Dub the tying silk with three parts olive-brown seal's fur to one part fluorescent pink seal's fur, well mixed, 3. Wind the dubbed silk body up the hook shank, aiming for a slightly tapered effect at either end, 4. Stretch the PVC over the top of the body, 5, and tie in, cutting off the waste. Complete with a whip finish.

Oh yes, the name Gerroff. Brian Clarke was fishing with Goddard, trying out the fly. The venue had many small trout and Clarke spent much time in inviting these small fish to release the nymph, aided by cries of 'get off'. As the day wore on it became blurred into a heart-felt cry of 'gerroff'. I trust that this explanation will prevent some future historian from coming up with a fancy, and highly improbable, answer!

Grey Nymph

Collyer

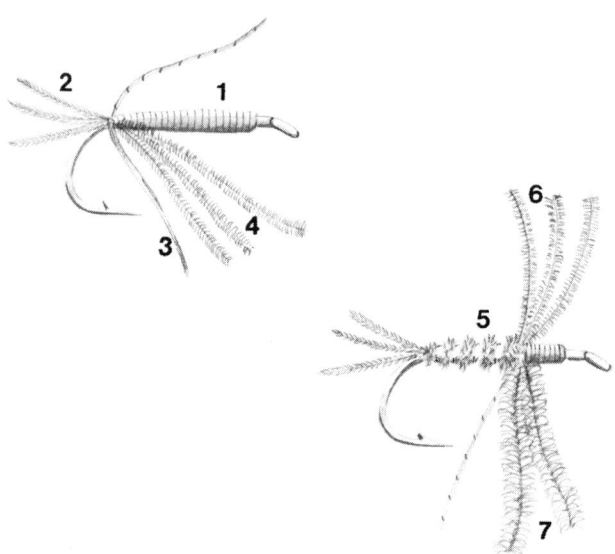

David Collyer is a very well known contributor to the fly-fishing press, in particular to the monthly magazine *Angling* wherein he regularly describes the tying of a particular pattern. The flies he describes are all good trout takers, though none more so than one of his original patterns, the Grey Nymph.

This artificial fly is one of a trilogy of nymphs devised by this master tyer, the Grey, the Green and the Brown Nymph, all accepted as most effective attractors, not only of many trout but of many large trout.

Though basically a stillwater pattern I know many fly-fishers who have used it with excellent effect on streams and rivers throughout the country. The river patterns need to be tied on smaller hooks than the reservoir patterns. I find hook size 14 quite admirable, but here we shall describe the original stillwater pattern.

The hook size is number 10. Take the black tying silk, 1,

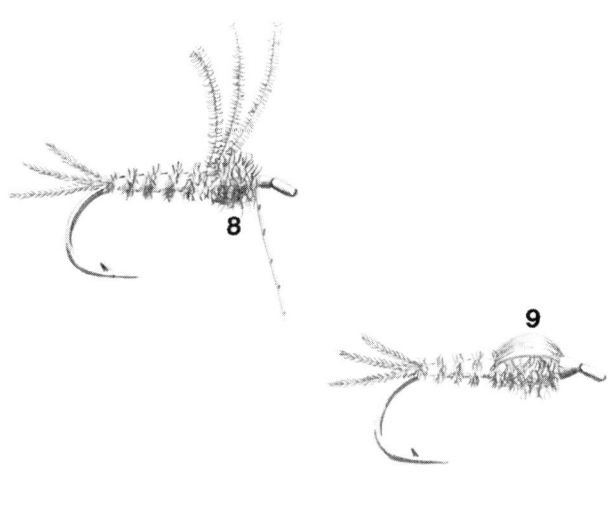

down the shank of the hook in close even turns. At the
bend of the hook tie in three natural coloured heron herls
from the primary wing feather as tails, 2, leaving the waste
ends trailing forward towards the eye of the hook. Now tie
in a length of oval silver tinsel for the rib, 3. Take the
heron herl ends, 4, and twist together, winding to form the
abdomen, followed by open turns of the ribbing material,
5. Tie in and remove the waste end of the ribbing but leave
the waste ends of the heron herls pointing over the eye and
on top of the hook shank.

Tie in two lengths of natural ostrich herl, grey with white
tips to the fibres, 7, and wind back and forth to create the
pronounced thorax, 8. Secure with a half-hitch and remove
the waste ends. Now bring the heron herls over the thorax
to form the wing cases, 9. Complete the fly with a carefully-tapered whip-finish.

Longhorns

Walker

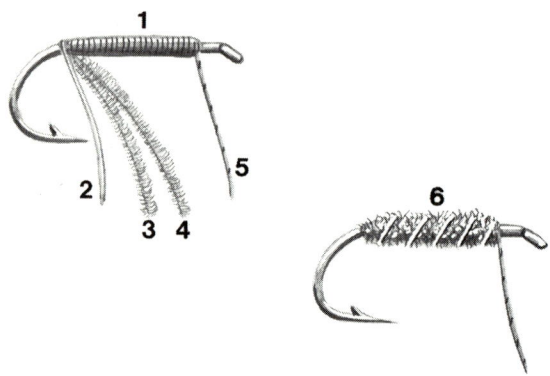

A first class pattern that is the brainchild of Richard Walker, who knows more than most anglers about the taking of stillwater trout. His regular series of fly patterns in the magazine *Trout and Salmon* has been instrumental in bringing to a wide audience a selection of fly patterns many of which have moved on to become firm favourites of the flyfishing world.

One of his most popular patterns has been his dressing of the Longhorn series, Walker's imitation of the various *Trichoptera* on their way up to the surface to hatch. Various body colour combinations are successful, though I have had considerable success with the original pattern here described.

The hook size is generally between 8 and 12. Personally I have found it most effective between 10 and 12. Start the primrose tying silk, 1, from some little distance behind the eye and wind in close even turns down to the bend of the hook. Carefully tie in a length of gold thread that will act

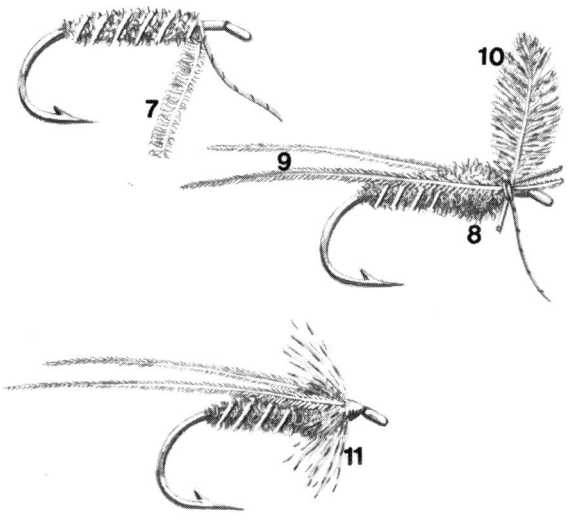

as the ribbing material, 2. Follow this with strands of ostrich herl dyed pale blue-green or amber, 3 & 4, then wind the tying silk back up the shank, 5. Twist together the ostrich herls and wind them back and forth up and down the shank to create a plump body. Wind the ribbing material of gold thread over the body in wide turns and secure. Cut off waste body and ribbing material, 6.

Tie in a length of ostrich herl dyed sepia or a chestnut colour, 7, and wind as the thorax, 8. Now tie in two fibres of cock pheasant tail feather fibre, as dark as possible, and tie them in so that they extend over the body for twice its length, 9.

Select a well-marked brown partridge feather and tie in over the roots of the pheasant tail fibres, 10. Wind the partridge hackle round the hook shank, two turns, and secure, 11. Complete the fly with a well-varnished whip-finish. (N.B. A later development is the use of dyed white rabbit fur in place of ostrich herl.)

Killer Bug

Sawyer

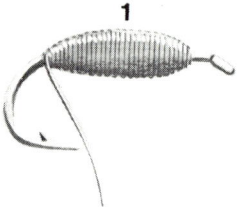

Some may consider that I am stretching the term 'nymph' to its ultimate by including in this book Frank Sawyer's excellent pattern the Killer Bug. I remain unrepentant, with head unbowed, for I have found it to be a most deadly pattern when hunting the prime grayling of the late autumn chalk streams. Any pattern that extends the fly-fishing season into the winter must be worthy of inclusion and I commend it to your attention.

Not only the grayling find it quite irresistible, for many experienced flyfishers have discovered the worth of this simple pattern when cast before the browns and rainbows of stream and still water. Even salmon have been known to look upon it as manna from heaven and so I do think it can be claimed to be a good all-round artificial.

That it is a Sawyer pattern is sufficient recommendation for you to give it a cast. The tying is quite easy and should be within the capabilities of the tyro fly dresser.

The hook can be between 9 and 12 for lake trout, 7 and 4

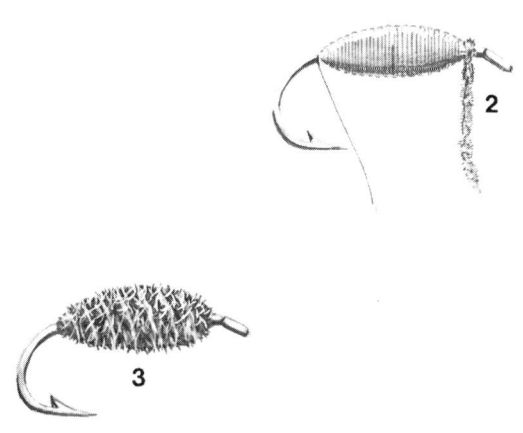

for salmon and 12 to 14 for the streams and rivers of this country. Having placed the hook in the vice proceed to wind copper or gold coloured fuse-wire round the hook shank, backwards and forwards, the while building up a tapered shape as shown, 1. Ensure that your winding activities end at the eye of the hook and there tie in a length of Chadwick's wool number 477. Sawyer is most insistent about the shade. Having tied this in, 2, take the fuse-wire back down the hook to the bend.

Now wind the Chadwick wool backwards and forwards over the fuse-wire underbody, ending up at the bend of the hook. Secure with the fuse-wire and cut off the waste. Now wind the wire in open turns up the body to secure the wool, 3. Carefully fashion a whip-finish with the fuse-wire, not easy, and drench with varnish. Cut off waste.

Unkind folk have said it is a perfect representation of *Vulgaris maggotia* but whatever it may look like it really does take fish.

Phantom Larvae

Collyer

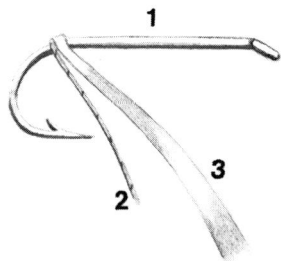

Another of David Collyer's well-thought-out patterns, this time a design that was commissioned by that well-known personality Alex Behrendt who has made such an outstanding success of the famous Two Lakes fishery in Hampshire.

Behrendt asked Collyer if the latter would devise an artificial that would equate to the almost transparent species *Chaoborus* in the larval stage. Not an easy thing to imitate and Collyer gave it much thought. The end result is here described. The pattern has achieved success not only at Two Lakes but on many other waters besides, indicating its worth in no uncertain manner.

The type of hook used is most important. A size 14 long shank but it must also be a silvered hook, not the usual blue-black. The translucent effect relies upon the silvered shank, rather like the white painted hook shanks specified

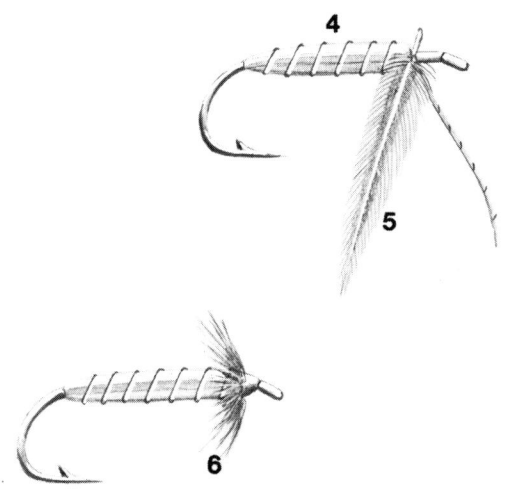

by Dunne in his book *Sunshine and the Dry Fly*, 1924.

To tie the pattern is not difficult. Let the hook shank remain clear of tying silk, 1, the silk being started at the bend of the hook and to be of well waxed olive, 2. Tie in a length of clear polythene, approximately one-sixteenth of an inch wide, 3.

Wind the polythene, maintaining a good tension on the material, up and down the hook shank, finishing at the eye end of the hook. Now wind the waxed silk over the polythene in wide turns of ribbing, 4. Tie in and cut off the waste polythene.

Now tie in a sharp badger cock hackle of short fibre, or a soft badger hen hackle, 5, in front of the body. Wind the hackle for two or three turns hard up against the body material, 6. Twitch off the waste hackle stalk and complete the pattern with a neat, well-varnished whip-finish.

Black Marabou Pupa

Price

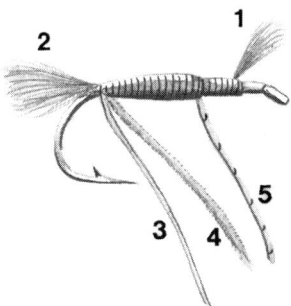

Yet another pattern devised by that excellent flydresser Taff Price, a man who has spent a great deal of time and effort transposing his entomological observations into very effective artificials for stillwater and stream angling.

The Black Marabou Pupa first saw the light of day some four or five years ago when Price formed the opinion that most pupal buzzer patterns lacked the movement of the natural species and he set about the task of using materials that would create this lifelike movement.

The method he finally adopted was the same as for his previously recorded bloodworm dressing, namely the use of marabou fronds, for he maintains that the slightest movement of the rod tip or line will activate the tail half of the dressing, causing it to wiggle and shimmer.

The hook size is generally between 16 and 10. Start the black tying silk down the hook and after but a few turns tie in a bunch of soft white hackle fibres to represent the fila-

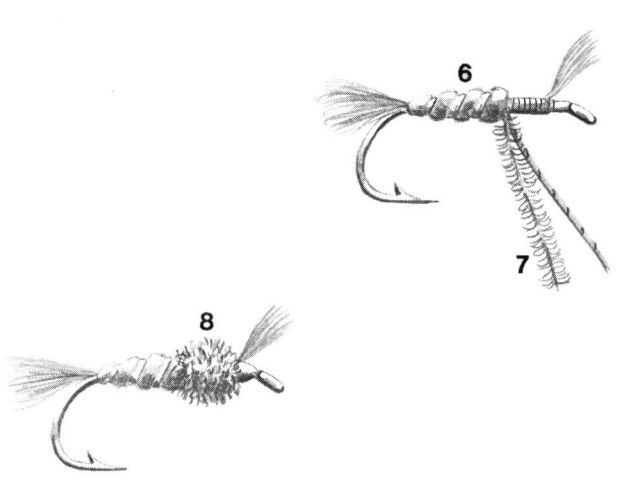

ments, 1. Continue the silk down the shank in close even turns to the bend and there tie in a tuft of black marabou feather frond, 2. This is in effect an extension of the body rather than a tail or whisk. Follow the marabou with a length of oval silver ribbing wire, 3, closely followed by a length of black floss silk, 4.

Wind the tying silk back up the shank to the point shown, 5, followed by the black floss silk and the ribbing wire, the latter in open turns, 6. Tie off the body material and rib waste and remove the excess.

Tie in a length of peacock herl, 7, and wind in the usual plumpish shape of a thorax, 8. Complete the artificial with a neatly-tapered whip-finish and varnish the head.

The pattern can also be tied in a range of colours; orange, olive, green, brown, etc. Fished during a buzzer hatch, just below the surface and very very slowly, the fly can be most effective.

Plain Brown Nymph

Nice

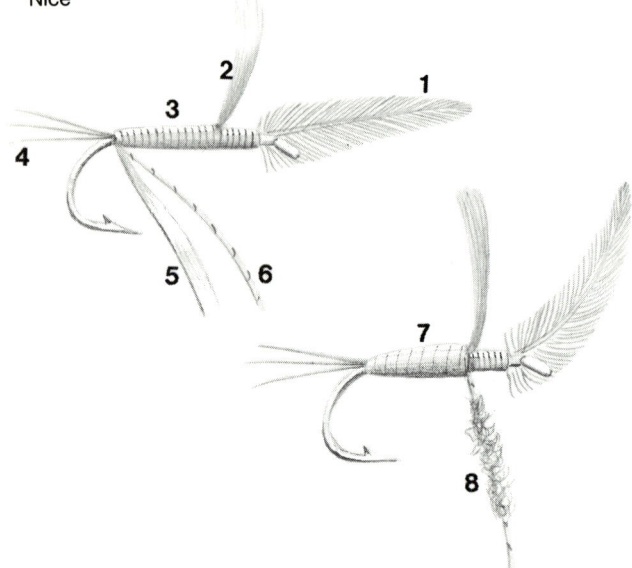

A further example of James Nice's carefully worked out patterns. This artificial was designed following his study of the various olive nymphs that he observed in the stomach contents of many Devon trout. Devised in the first instance for stream fishing it has also proved its worth on many stillwater locations.

The amount of time Nice spends on experimentation may be seen from the following letter extract: 'At first for the winding of the body material I tried balloon rubber, but experience has shown that opaque PVC strip, such as used for baby-pants, is a much better material and is far more durable. I have tried various ribbing mediums under the PVC and have also tried fluorescents under the body but experience and many recorded experiments have proved such features do not improve the fish-taking ability of the pattern'. Such is the detailed analysis top class tyers carry out.

Again Nice is most insistent in the number of turns of tying silk, etc., when using a standard hook size 14. Start the silk down the hook shank for six turns, tying in a

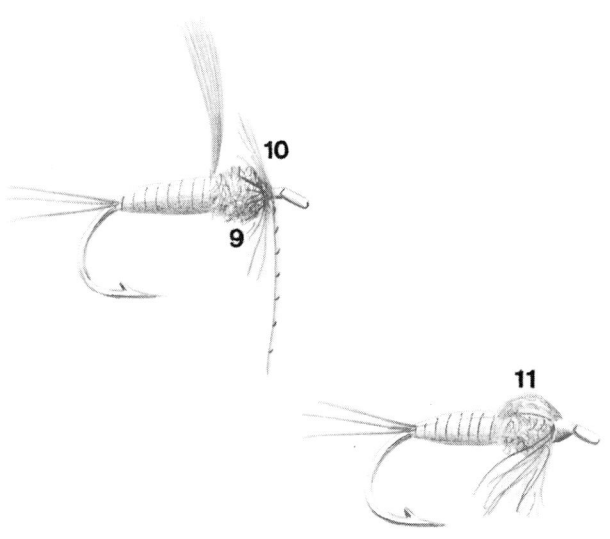

brown-olive hackle by the stalk, 1. Continue the silk down the shank for a further twelve turns, trapping the hackle stalk underneath. Cut off the waste stalk. Now tie in brownish feather fibres for the wing-case material, 2, with the waste ends pointing towards the bend. Continue the silk in close turns, usually six, at which point remove the waste ends of the feather fibres and complete the windings to the bend, 3, at the same time tying in three brown-olive cock hackle fibres for the tail, 4.

Now tie in a length of opaque PVC dyed brown, 5. Return the silk, 6, in close even turns up the shank to one turn in front of the wing case material. Wind on the PVC in even turns, just overlapping, to form the abdomen, 7. Tie in and cut off waste end. Dub the tying silk with brownish fur, or brown herls, 8, and wind to form the thorax, 9. Wind the hackle in front of the thorax for a maximum of two turns, 10, then pull the wing case material over the top of the thorax, dividing the hackle fibres on each side, 11. Complete with a well-varnished, neatly-tapered head.

Shrimp

Overfield

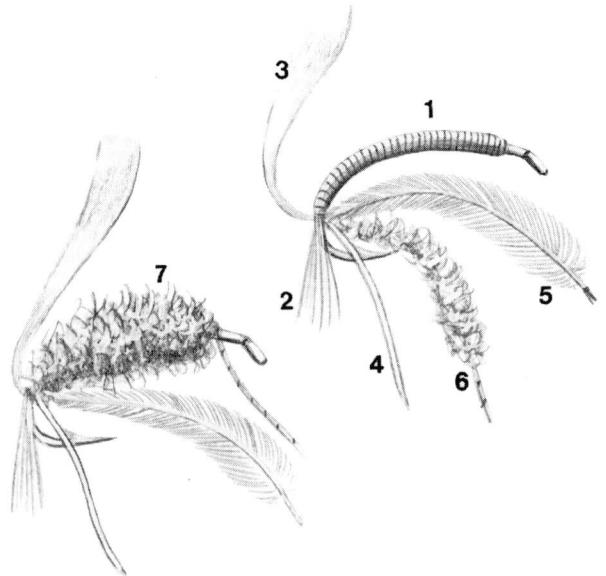

At the risk of appearing awfully egotistical I intend to include one of my own flies in this series. My excuse is that I have had considerable success with this artificial and would hope fellow flyfishers would care to try it.

The immediate difference between this pattern and many others is the shape of the hook. I tie it on hooks ranging between 14 and 10 but they all have two things in common; they are of the wide-gape variety and I very carefully bend the shank to obtain a curvature, simulating the shape of the natural. Care must be taken in the bending of the hook for if it is well tempered it is not the easiest thing to manipulate.

The Shrimp can be weighted if desired by the use of fine copper wire wound onto the shank, ideally creating a hump-backed effect.

The tying silk is of a pale-olive colour and is wound down the shank in close even turns to a position well round the bend, 1. There tie in a bunch of light-ginger cock hackle fibres with a pronounced downward slant, 2.

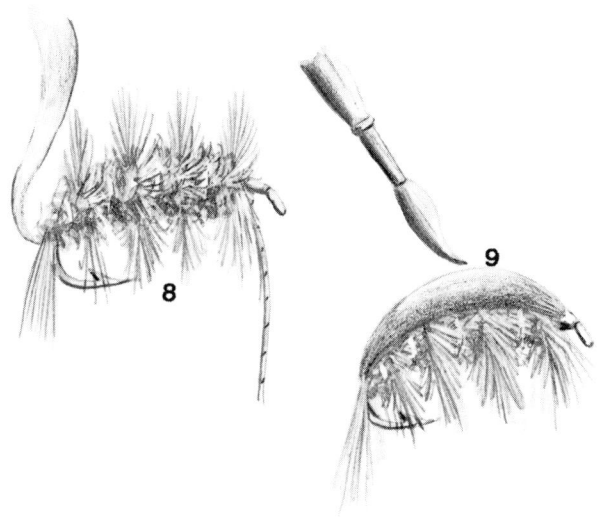

31

Follow these with a length of light-olive PVC strip, 3, and a length of silver wire for the ribbing, 4.

Now tie in a short fibred light-ginger cock hackle by the point, 5, and having re-waxed the tying silk dub it with a mixture of hare's ear fur and pale orange seal's fur, very well mixed, 6.

Wind the dubbed silk backwards and forwards over the hook shank to create a hump-backed effect, 7, remembering to end the operation at the eye end of the hook. Secure with a half-hitch and remove waste dubbing. Wind the hackle in open turns over the dubbed body, followed by the silver wire ribbing to secure the hackle in place. Tie off and remove waste hackle and rib, 8. Take the PVC and, carefully separating the hackle fibres to either side, draw it over the top of the body, maintaining an even tension. Tie down and complete the fly with a whip-finish, removing the waste PVC.

With a fine paint-brush carefully coat the PVC strip with clear varnish, 9.

Spurwing Nymph

Waites

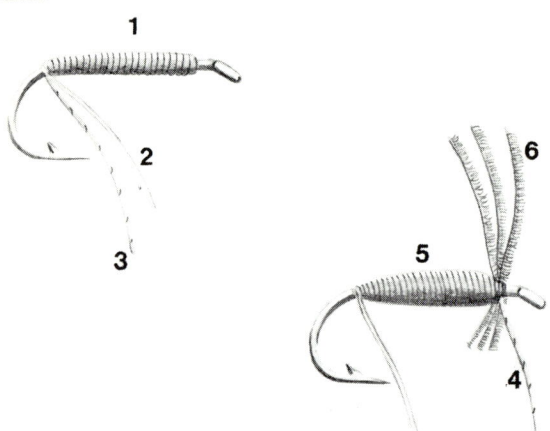

This is a very simple though highly effective dressing. If I were to be confined to a very few nymphal patterns then this would be in the first three.

I first saw it being used some six years ago on the Driffield chalk stream and I do feel the manner of our meeting is worth recounting. My wife and I were fishing the Island beat of that particular stream on a very hot cloudless morning in August. A respectable number of duns had floated down during the morning but the trout were intent on nymphs and nymphs only. We had tried most of the patterns in our respective fly boxes and while one could see the trout move over to inspect the offering they did not take.

At lunchtime we were frustrated and very hot and opted to have our picnic lunch. The head river-keeper, Tony Waites, joined us and we talked of the morning. From his lapel he produced a small grey nymph tied on a number 16 hook. Taking my rod he tied it on and crouched at the bankside. We watched with interest. He moved no more than fifteen yards and within half an hour hooked and released five brown trout! His stream craft obviously played a part in the exercise, as did his superb eyesight, for he can see the slight flash of white when the trout opens and closes its mouth. But he would agree that I am not exactly lacking

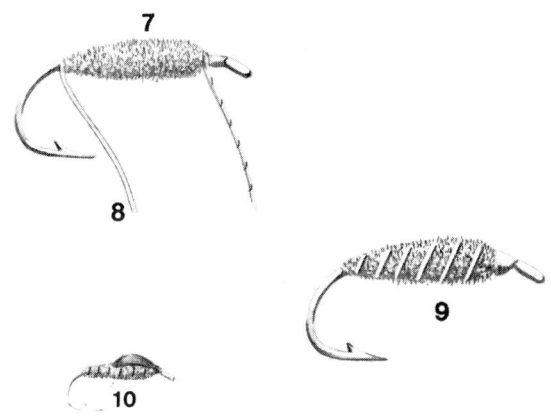

in fish-taking ability and that the artificial nymph had quite a lot to do with this astonishing performance. While we finished lunch he tied up a few examples for us and left us to our own devices.

In true storybook manner we seemed to have found the magic touchstone and that afternoon, and many days since, have given us cause to bless the day we discovered this particular pattern. Now whenever we are in doubt we put up a Waites Spurwing and fish with considerable confidence.

The hook size can be between 18 and 14, with 16 the usual favourite. Carefully wind an underbody of fuse-wire, 1, followed by grey tying silk. Do not worry about forming a thorax shape with the fuse-wire, for the lack of it does not detract from the pattern. Tie in a length of fine silver fuse-wire, 2, then take the silk, 3, back up the hook to point 4, creating a tapered shape, 5. Now tie in two or three natural heron herl fibres, 6. Wind these backwards and forwards, 7, securing and removing the waste ends. Wind the fine silver wire, 8, in open turns up the body, 9. Complete with a whip-finish.

Alternatively one can include wing cases, again of heron herl, 10, in which case do not cut off the waste ends after stage 7 but double and re-double them in the usual manner.

Carrot Nymph

Skues

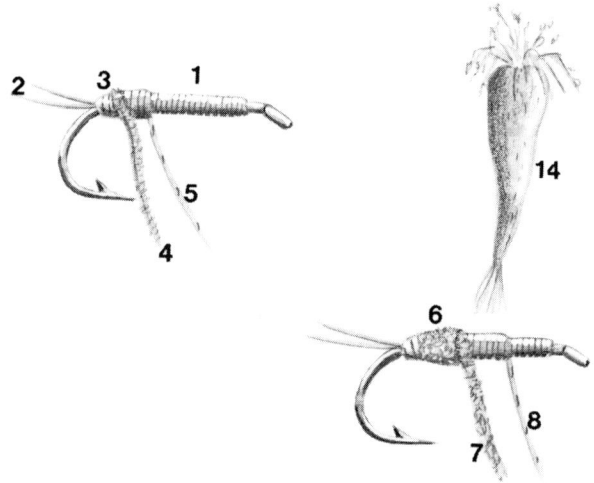

This must be the joke nymph of all time. G. E. M. Skues invented a mythical character called 'The Novice' who did all sorts of unspeakable things piscatorial on the banks of the Itchen and (in 1912) was responsible for the Carrot Nymph. That Skues intended the pattern to be taken in light vein is obvious. Little did the great man know that this 'vile creation' would one day be tried with great success on a stream far removed from his beloved Itchen.

Some years ago I had tied up a few examples of the Carrot Nymph, more as a whimsical exercise than with any serious thought of fishing such an odd creation. Came the day when my usual armoury of nymphal representations proved completely ineffective against the trout. I looked in my fly box and there rested five examples of the Carrot Nymph vile. Dare I? The river-keeper was away on another beat and so I felt reasonably confident of not being caught offering his precious brown trout such an unspeakable offering. I will not bore you with the details but the Novice's creation brought me three very fine trout. Do not embarrass me by asking what natural they thought it to represent. I suspect that even that fine entomologist John

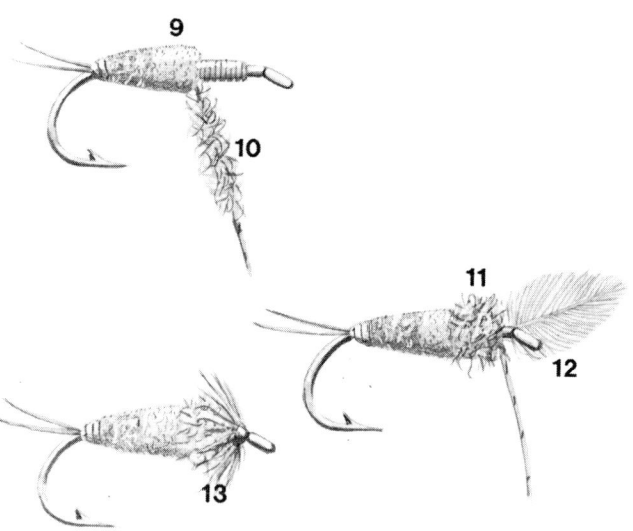

33

Goddard would be stumped for an answer. I do know that since that day other trout have fallen to the nymph that started life as a joke back in 1912. I only wish Skues were alive today for I am sure no-one would have appreciated the humour more than he.

For those who are brave enough to try it I will give the dressing. The hook size is 14 to 16. Take the primrose tying silk, 1, down to the bend. Tie in whisks of green parrot feather fibre, or substitute, 2. Wind the silk back up the hook shank for three or four turns to create a tag, 3. Tie in a length of pale yellow wool, 4. Take the tying silk in close turns up the shank, 5, covering the silk with the turns of wool, 6. Cut off the waste and tie in a length of hot-orange wool, 7. Wind the silk forward again, 8, followed by the orange wool, 9. Cut off the waste end. Dub the silk with greenish seal's fur, 10.

Now wind the seal's fur round the shank as a thorax, 11. Tie in a short fibred cock hackle dyed olive-green, 12. Wind the hackle, two turns maximum, round the hook and complete the fly, 13.

Aim to achieve the tapered effect of the natural, 14!!

Sedge Pupa

Goddard

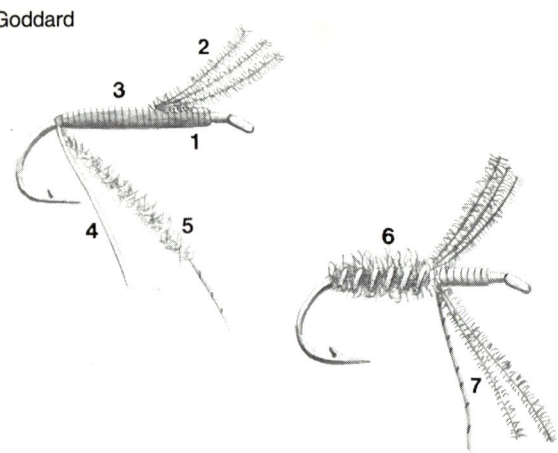

One of the best dressings to represent the pupae of a wide variety of sedge flies. Devised by that talented amateur entomologist John Goddard, who considers that the artificial is usually more productive from June or July onwards. Goddard also rightly points out that maximum success will be achieved during the emergence period of the natural insect, though many stillwater anglers have found excellent sport with this pupa fished throughout the season as a general pattern. However, when you see the adult winged sedges fluttering on the water surface that is really the time to tie on this pattern to the point of your leader. The inventor recommends two methods of fishing, both on a floating line; the slow sink-and-draw retrieve to imitate the natural swimming motion up to the surface, or the slow consistent retrieve to simulate the activity of the natural just below the surface. Of the two methods I have found the latter to be very effective when fishing the margins.

Tie the artificial in various shades to represent the most common colours of the naturals—the body material should be either olive-green, dark brown, orange or cream.

The hook size is between 10 and 12 of the long shank, wide gape type. Take the tying silk of a brown colour, 1, in close even turns down the hook shank for the length of the thorax. Tie in lengths of pale-brown condor herl, 2, –

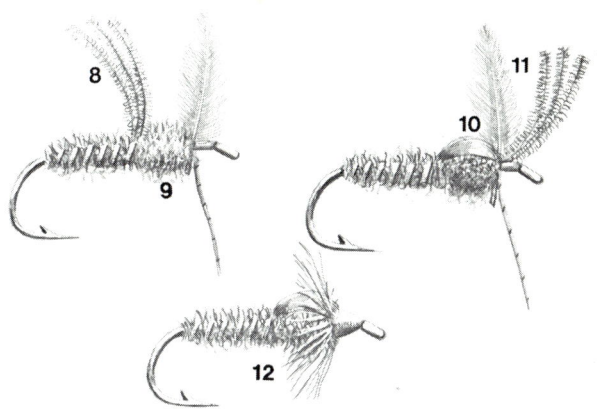

dyed turkey can be substituted for condor herls – trapping the waste ends under the silk as you continue to wind the silk down to the bend, 3. Tie in a length of narrow silver lurex, 4. Now re-wax the tying silk and dub with seal's fur of the appropriate body colour, 5. Wind the dubbed silk up the hook shank as far as the condor herls, followed by open turns of ribbing with the silver lurex, 6. Tie off and remove waste lurex and dubbing material. The seal's fur dubbing may be lightly covered with fluorescent floss of the same colour for greater effect, the whole being ribbed with the silver lurex. If you wish to do this the fluorescent materials should be tied in between stages 4 and 5.

Take the silk in front of the wing case material and tie in dark-brown condor herls, 7. Hold back the wing case material, 8, and wind on the condor herls, backwards and forwards to form the thorax, at the same time tying in a rusty-dun hen hackle, 9.

Bring the wing case material over the thorax, 10, then take the waste ends, 11, back over the thorax, having previously taken the silk to the rear in wide turns ready to secure the same waste ends. Take the material forward again and secure, removing the waste ends. Wind the hackle for a maximum of two turns, 12. Secure with a well-shaped whip-finish and varnish same.

The Persuader

Goddard

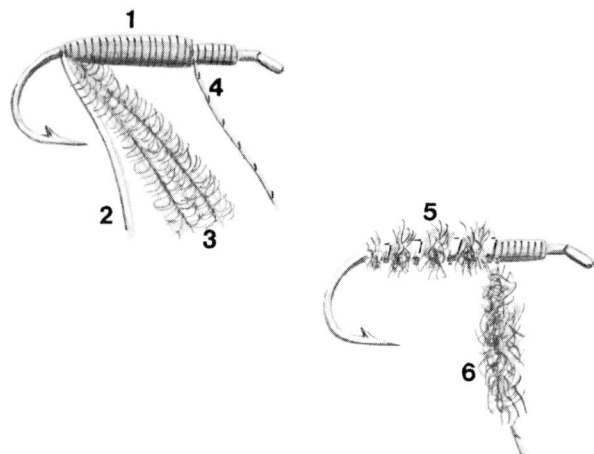

Another popular pattern from the vice of John Goddard, being quite simple to tie and yet a pattern that was the result of much careful investigation.

First tied in 1975 for the trout that inhabit the Hanningfield reservoir it has since that year proved its worth on many other stillwater fisheries. The design criteria were three-fold; it had to be of a fairly large size to attract trout from a considerable distance. The colour combination had to be attractive to the fish. It had to represent a range of natural nymphs or pupae.

Goddard spent many hours experimenting with various colour combinations and shapes, finally deciding that white ostrich herl was particularly translucent and therefore ideal for the body material. The seal's fur thorax had similar sparkling and translucent qualities. The overall shape was to be that of a sedge pupa.

The method of fishing this pattern is either on a sinking line, fished quite slowly as near to the bottom as possible, or alternatively retrieved at quite a fast rate just below the surface using either a floating or a sink-tip line fished in

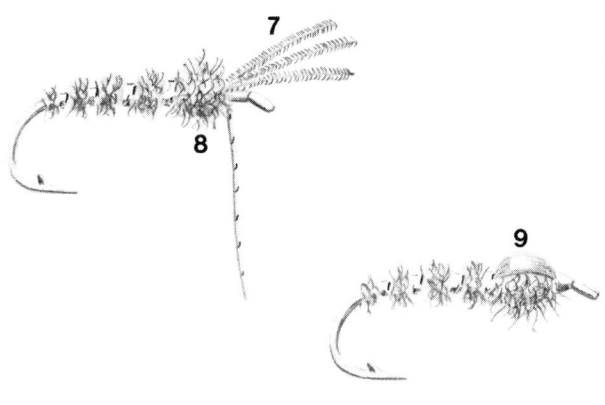

shallow water. It is occasionally useful in deep water if you have noticed trout rising in that area. Goddard tells me that the pattern has also accounted for a considerable number of trout when cast out into deep water, using a floating line, and allowed to sink slowly.

A very simple and quick pattern to tie. The hook is 8 or 10 long shank. Wind the orange tying silk, 1, down to the bend of the hook. At that point tie in a length of round silver tinsel, 2, for the ribbing, followed by five strands of white ostrich herl, 3. Now wind the tying silk back up the hook shank to a point, 4. Wind the ostrich herl up the hook shank and rib it with open turns of silver tinsel, 5. Cut off waste herl and ribbing.

Dub the tying silk with orange seal's fur, 6. At the same time tie in three strands of dyed dark-brown turkey fibres, 7. Now wind the dubbed silk to form the thorax, 8. Bring the silk in an open turn to the rear of the thorax and then pull the turkey fibres over the thorax to create wing pads. Repeat backwards and forwards and tie off, 9. Complete the pattern with a whip-finished head and varnish.

Water Tiger

Collyer

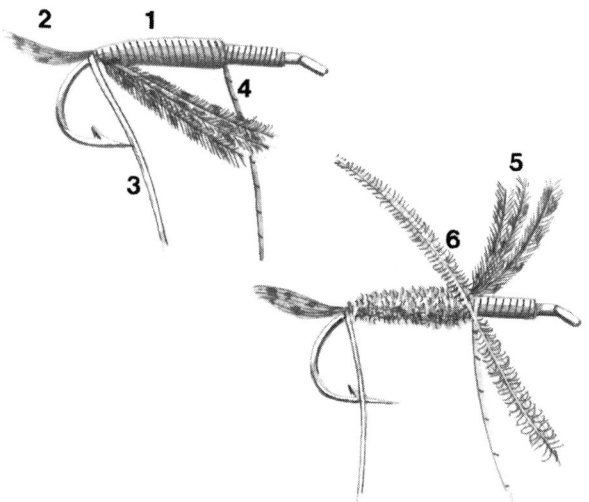

Having carried out numerous stomach contents examinations on trout taken from a wide variety of southern reservoirs David Collyer observed large food forms that he eventually identified as Water Tigers, the larvae of the Great Diving Beetle *Dytiscus marginalis*, a most aggressive creature that does not hesitate to attack very small fish. Collyer recalled that as a child when engaged on a newt catching expedition he often felt his fingers being nipped by the aptly named Water Tiger.

This creature generally frequents shallow water areas of lakes and reservoirs, the adult forms rising to the surface at sporadic intervals to take in air. The larvae are found on the bed, however, and therefore the artificial should also be fished on the bottom with short retrieves and lengthy pauses.

The natural larval form is quite large and would require a size 4 long-shank hook, but Collyer considers such a size would look most odd, and would not be allowed on some fisheries, and so he settled for size 10 long shank, at most a size 8. The trout do not seem to mind the diminution of their food.

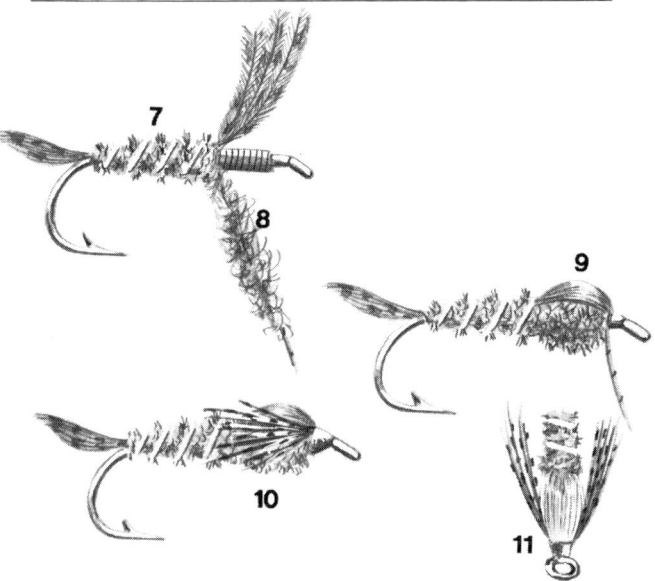

36

The tying is a little bit complicated, so pay close attention. Having secured the hook in the vice take the brown tying silk down the hook shank, 1, and at the tail tie in fibres of speckled turkey tail, 2, or sepia dyed condor herl, allowing the waste ends to lie forward. Now tie in a length of copper wire for the ribbing, 3. Wind the silk in close even turns back up the shank, 4. Take the body fibres and wind over the silk, still with the waste ends pointing forward, 5. Now tie in a single strand of bronze peacock herl, 6, and wind one end down to the tail. Secure with the copper wire and wind the latter up the herl in open turns as ribbing, 7. Twitch off the waste end at the tail.

Dub the tying silk with yellowish-olive wool or seal's fur, 8, and wind on to the hook shank in a pronounced thorax shape. Bring the waste ends of the body material over the thorax as the wing cases, 9. Now tie in on either side of the thorax two bunches of brown partridge feather fibres, 10, seen in plan view at 11.

Complete the pattern with a neat whip-finished head and a coat of varnish.

Blue Winged Olive

Jacobsen

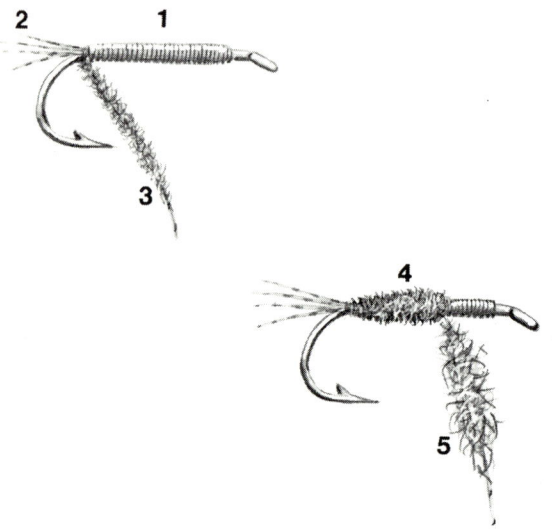

The second of Preben Torp Jacobsen's patterns to appear in this book, and a nymph that has its origins in Skues' version of the Blue Winged Olive.

Jacobsen confirms that G. E. M. Skues has been his spiritual mentor down the years via such books as *The Way of a Trout with a Fly* and *Nymph Fishing for Chalk Stream Trout*, etc., but those of us who have had the distinct pleasure of knowing Preben consider that the pupil has caught up with the master and in some directions may even have gone ahead. Certainly in the area of material manipulation I would put Jacobsen in the top five flytyers of the world.

This version of the nymph of the Blue Winged Olive comes high on the Dane's list of favourite patterns and I would feel rather lost if a few such artificials were not regular inhabitants of my own fly box.

Jacobsen's nymph incorporates a material that one does not find very often in modern fly patterns, cow hair. The tyers of the Golden Age certainly recognized its value and

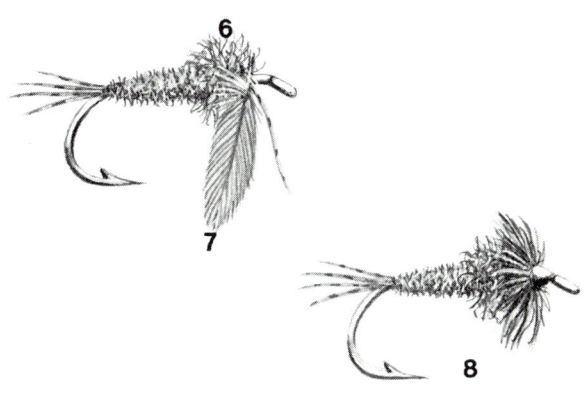

most of the classic books of the past include cow hair in the statutory list of furs and hairs that one should carry in the dubbing bag.

The tying is generally done on a size 14 hook. Take the hot-orange tying silk, 1, down the hook shank in close even turns to the bend. Tie in three or four partridge hackle fibres, 2, and then dub the tying silk with a very small amount of otter's fur spun onto the silk very tightly, 3.

Wind the dubbed silk up the hook shank to form the abdomen, 4. Remove the excess body dubbing and re-dub with blood-red (dark brownish-red) cow's hair, 5. Wind the cow's hair onto the hook and form a pronounced thorax, 6.

Tie in a very small dark-blue hen hackle, 7, and wind in front of the thorax for a maximum of three turns, 8. Complete the nymph with a neatly-tapered, well-varnished whip-finish.

This really is a most effective little nymph and I do urge you to try it.

Smut

Mottram

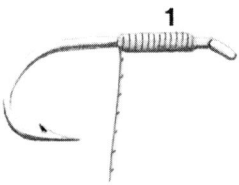

While the birth of the present day nymphal artificial form may quite correctly be laid at the door of G. E. M. Skues we should not forget that others had a positive hand in the development. One man in particular who devoted a lot of time and effort to this subject was J. C. Mottram who in his book *Fly Fishing: Some New Arts and Mysteries*, circa 1920, set forth a most convincing argument for the use of the representational nymph in upstream fishing. That Mottram changed his mind and did in fact become one of Skues' greatest opponents over the use of the nymph for chalk-stream trout does not alter the fact that he came up with some very good nymphal patterns. One of the best to my mind is the Smut, which if one thinks about it may not in fact be a true nymph. I do take the broad view, however, that as it is fished upstream and just below the surface it may well qualify for my own particular definition. Purists must forgive me, though before they do I suggest they try

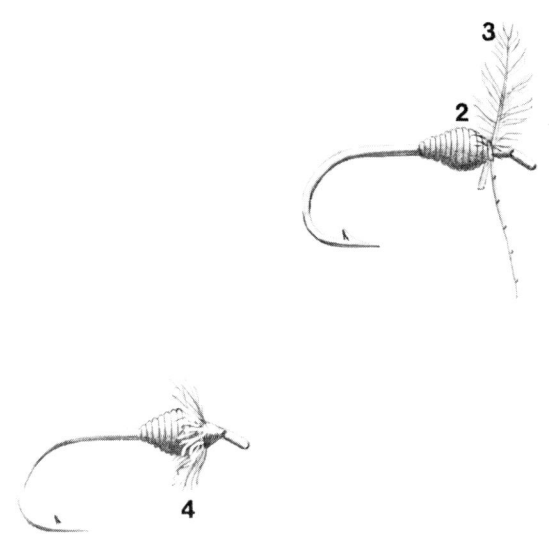

the pattern, especially on streams, whereupon they may discover there is nothing to forgive.

Few patterns are so simple in the tying. The hook size should be between 14 and 16. Start the tying silk of black floss, 1, down the hook shank and continue to wind for no more than half the length of the shank. Now return the silk on close even turns back up the hook, the while creating a thorax type shape, 2. In length the body should equate to the typical Clyde-style fly. Now tie in a small starling breast feather, 3, and wind on only one turn at most to form the hackle, 4. Whip-finish the head into a neat taper and varnish with clear varnish.

My favourite method of fishing the Smut is to grease the leader to within two inches of the point, casting upstream to obviously nymphing trout with confidence. You may well be surprised but most certainly not disappointed.

Dragon Fly Nymph

Price

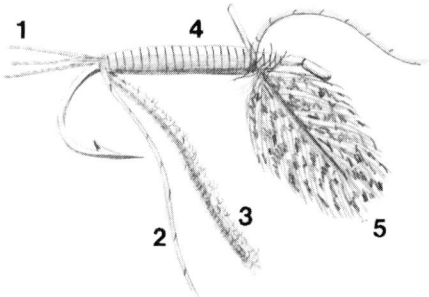

Here is a fly pattern that has proved its worth on those hot windless days described by its inventor Taff Price as 'those sods of days' when absolutely nothing moves on the surface of still waters.

The pattern has the distinct advantage of being easily tied from readily obtainable materials, for Price comes down very hard on those flies that demand the use of 'the third primary feather of the lesser spotted dung hawk, taken only when the moon is in its third quarter, and then only with gloved hands'. No, the Dragon Fly Nymph is made from those materials that are generally found in all flytyers' cabinets. Despite its humble origins it is a highly effective pattern.

The hook size is between 12 and 8, long shank. The brown tying silk is taken in close even turns down the shank to the bend, at which point three stiff goose feather fibres, dyed brown, are tied in for the tails, 1. These should

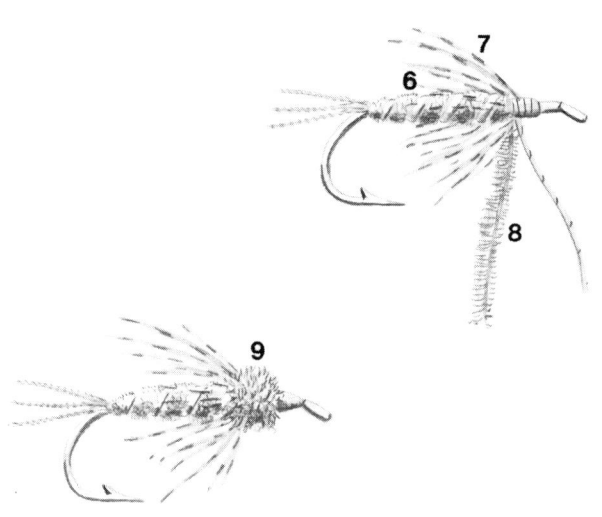

be quite short. Next tie in a length of fluorescent green silk as the ribbing material, 2, followed by the length of brown wool, 3, that will form the abdomen.

Now return the silk in close turns up the hook shank, 4, and tie in a brown partridge feather, 5. Wind the brown wool in close turns up the hook, followed by open turns of the fluorescent green silk as the rib. Secure the wool and rib and remove the waste ends, 6.

Wind the brown partridge hackle round the hook, ensuring a pronounced backward slant to the fibres, 7, and then tie in a length of peacock herl, 8. Carefully wind the herl round the hook shank to form a neatly shaped head, 9. Now finish off the artificial with a carefully tapered well varnished head.

For best results, according to the pattern's inventor, fish it very slowly deep sunk along the bottom of the lake or reservoir, and await surprising results!

Blue Winged Olive Nymph

Skues

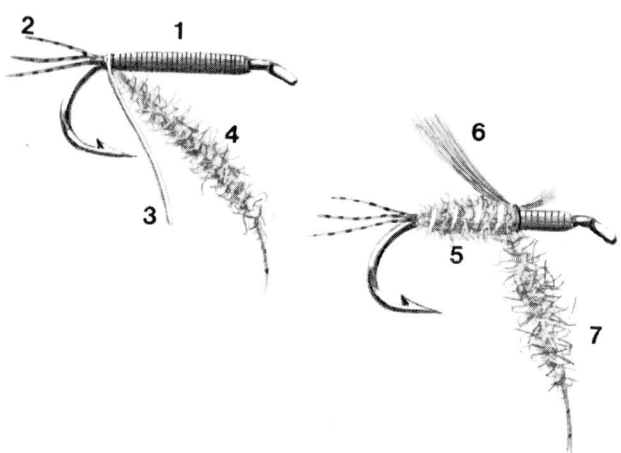

Yet another top class pattern from the stable of that master nymph tyer G. E. M. Skues, who wandered the banks of the river Itchen for so very many years. Skues was dead set against the grayling inhabiting his beloved trout stream. Like so many chalk-stream flyfishers he had very little regard for the species, though I must confess I do not consider them vermin. I would be happy to argue the merit of the grayling into the wee small hours, but I often think that I am a lone voice crying in the wilderness. My regard for the fish is mainly because it extends the flyfishing activity into the Christmas period. I must qualify that statement by saying that too many grayling will ruin a decent stream, but a yearly cull by electrically fishing the water can soon rectify that matter, leaving only the best and largest grayling to provide sport for the fly during the fence months when I wander the banks of the chalk stream, usually in solitude. The charm of the chalk stream is not to my mind confined to the high days of the trouting season, for withered grasses and hoar frost also have appeal, especially when a grayling takes the fly.

That Skues devised this nymphal pattern of the Blue Winged Olive especially for the taking of grayling makes me wonder about the old boy's professed loathing of the

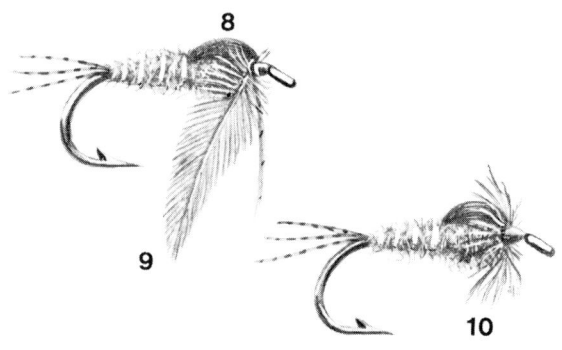

species. Whatever his reason for tying it I have always been grateful to him for devising a very good fly.

The hook size is between 14 and 16. Wind the pale orange tying silk down the hook in close even turns, 1, to the bend of the hook and there tie in three short strands of brown mallard feather fibre for the tail whisks, 2. Follow this with a length of fine gold ribbing wire, 3. Dub the waxed tying silk with a mixture of brown-olive seal's fur mixed with blue cat's fur dyed to a yellow-olive in picric acid, 4. Wind the silk and body material onto the shank, followed by the ribbing wire, 5. Now tie in a slip of heron's herls for the wing cases, 6. Re-dub the silk with the fur, 7, and wind to make a pronounced thorax. Bring the herls over the top of the thorax and tie in, 8, cutting off the waste ends. Tie in a small dark-olive cock hackle with a freckled appearance, 9, and wind onto the hook shank immediately in front of the thorax, with a maximum of three turns, 10. Complete the fly with a small neat whip-finished head and varnish carefully.

Go to the river or stream when the trout season is over and fish this excellent nymph with confidence to the grayling. You will not be disappointed.

Chew Nymph

Clegg

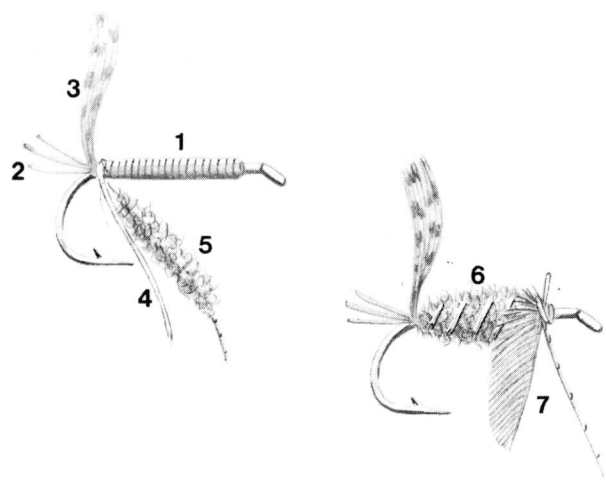

Another example of Thomas Clegg's carefully designed patterns, showing his restrained though highly effective use of fluorescent materials.

Some years ago fluorescents were the 'in thing'. Everyone seemed to be coming up with fly patterns that included this seemingly magical material. Interest then seemed to wane and I sincerely believe that it was because so many flytyers, professional and amateur, tended to over-egg the pudding, mistakenly believing that the more fluorescent material used the better the trout taking ability of the fly. They were wrong, completely wrong, for it is the restrained use of fluorescents that brings success.

Another top-class flytyer who knows how to use this 'wonder material' in small amounts is James Nice of Sidmouth, Devon, and his tyings are much sought after by discerning flyfishers, but it is to Thomas Clegg that we owe a debt of gratitude for producing a good book on the material and its uses, for those who would know more. His book *The Truth About Fluorescents* is obtainable from Tom Saville Ltd., of Nottingham.

Clegg's version of the well known nymph of the Chew

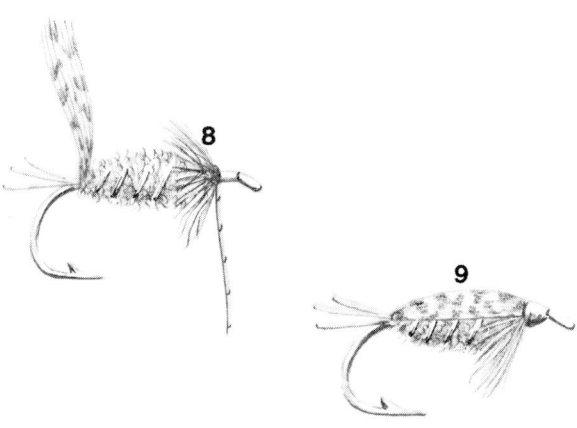

reservoir has achieved considerable popularity and certainly deserves its place in this small book. I am sure it is worth a trial on all stillwater fisheries throughout the country.

The hook is generally between sizes 10 and 8. Take the red tying silk, 1, down the hook shank in close even turns to the bend of the hook shank. There tie in three short lengths of Neon-Magenta DRF (Depth Ray Fire) filaments, 2, or three short lengths of DRF floss of the same colour. Now tie in a bunch of mottled turkey feather fibres, 3, that will provide the back to the pattern. Follow these with a length of Neon-Magenta floss, 4.

Dub the tying silk with mole's fur, 5, and wind the silk and fur up the hook shank to form a bulky body, this being ribbed with the DRF floss, 6.

Tie in a folded brown hen hackle, 7, and wind it in front of the body, 8. Draw the turkey fibres over the entire body and press down the hackle on either side, 9. Complete the nymph with a carefully-tapered and varnished whip-finished head.

Hatching Midge Pupa

Goddard

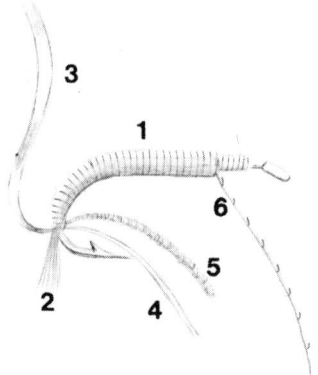

One of the earliest patterns devised by John Goddard specifically to represent the pupae of the *Chironomidae* of still water hanging motionless in the surface film prior to hatching into the adult winged buzzer.

As the artificial was designed to hang in the surface film it was vitally important that the pattern followed the old adage of 'exact representation', or as near as it is possible to get with fur and feather.

Working from photographic enlargements of the naturals at that particular stage in their life cycle, Goddard eventually evolved a most lifelike pattern that has been one of his firm favourites over many a year.

He usually ties them on straight-eyed hooks, which allows him to mount two or three directly to the leader with three-foot spacings. This arrangement, fished on a floating line with a lightly greased leader, allows the flies to hang head up in the surface film.

Cast out to trout during a buzzer rise this pattern, in various colours to imitate the natural of the moment, is quite deadly. It is not the easiest of patterns to tie for it is important that its shape and effect must bear close resemblance to the natural at that stationary stage in its life cycle.

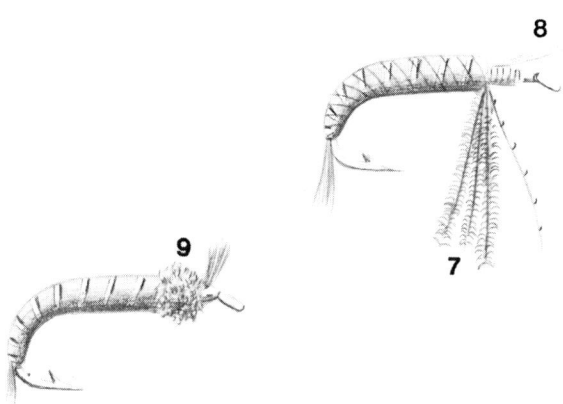

42

The hook should be between 10 and 14. Take the tying silk, 1, which should be the same colour as the body, down the hook shank and two-thirds of the way round the bend of the hook. Tie in several strands of white nylon filament to form the tracheal gills, 2, extending for approximately one-eighth of an inch. Now tie in a length of PVC strip, 3, followed by a length of silver lurex for the ribbing material, 4.

Next tie in a length of marabou floss silk of the appropriate body colour, 5, either black, brown, red or green. A strand or two of fluorescent wool can also be included if desired.

Wind the tying silk back up the body to 6, followed by the wool underbody. Now rib the wool with the lurex and then cover all with the PVC strip, removing all surplus materials. Tie in three strands of peacock herl or brown dyed turkey feather fibre, 7, also short strands of white fluorescent wool for the head filaments, 8.

Wind the thorax, 9, with the peacock herl and lift up the white wool filaments while finishing off the pattern with a neat whip-finish. Cut off the excess wool fibres.

Large Red Midge Pupa

Nice

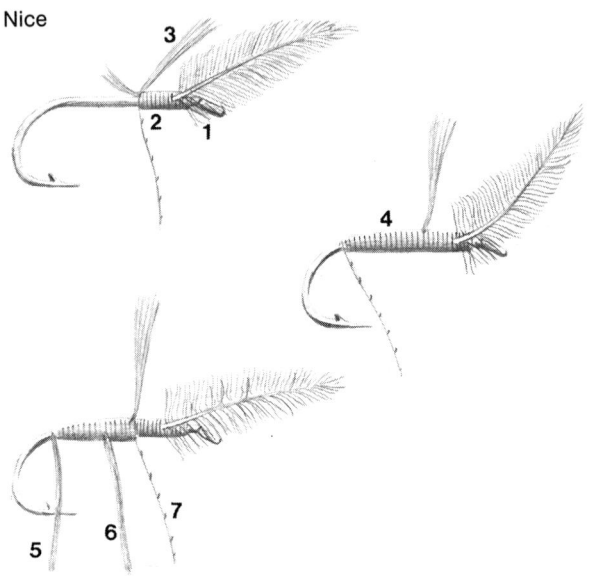

A further interesting pattern devised by James Nice of Sidmouth showing a most effective way to use fluorescent materials in the build-up of the abdomen while showing colour gradations.

A top class pattern to be fished in the usual buzzer style. My main reason for the inclusion of this artificial is to show just how an observant and skilled flytyer can use fluorescents to good effect. The tying is not easy but I do urge you to try Nice's method of body build-up, not only on this pattern but on any fly that you tie where you have a natural model in front of you from which to note the colour gradations.

Again Nice's instructions are very precise. The hook should be size 12 standard Limerick either D/E or straight eye. The pale brown tying silk is taken down the hook shank for six turns at which stage a buff, or palest ginger, hackle is tied in, 1. Tie down the stalk with a further ten turns of silk, 2. Cut off the residue of the hackle stalk. Tie in a bunch of the palest cinnamon feather fibres, 3. Whip down the butt ends at the same time cutting them at an angle with a sharp pair of scissors so that as you continue

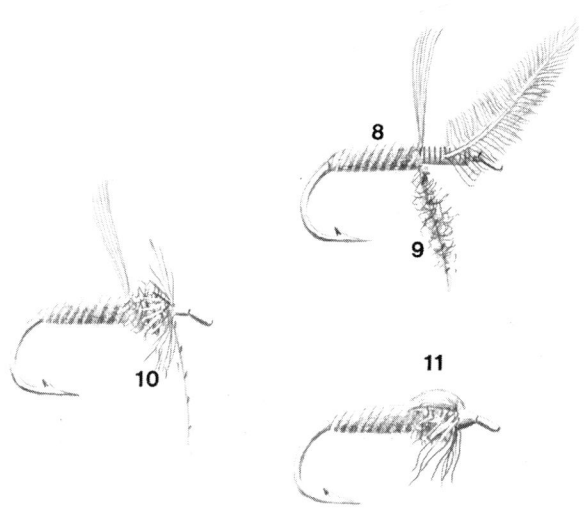

to wind the silk towards the hook the underlying fibres help to produce a tapered abdomen, 4.

Tie in a length of orange DFM floss, 5, making sure it is very secure, for this floss is quite slippery. Return the silk half way up the abdomen and there tie in a length of red DFM floss, 6. Now wind the silk one turn past the wing case material, 7.

Wind the red DFM up the hook and secure just in front of wing case fibres. Now wind the orange DFM floss up from the bend to the wing case. Tie off and remove waste, 8. The colour effect is quite excellent with the red DFM being hazily visible under the orange.

Dub the tying silk with buff coloured hare's poll fur, 9, and wind backwards and forwards to imitate the thorax. Wind the silk for two turns in front of the hackle then wind the hackle for two turns at most. Secure with the silk and remove waste end, 10.

Bring the wing case feather fibres over the hackle and secure, 11, tapering off the waste material under a four-turn whip-finish. Varnish the head with clear varnish.

Golden Shrimp

Price

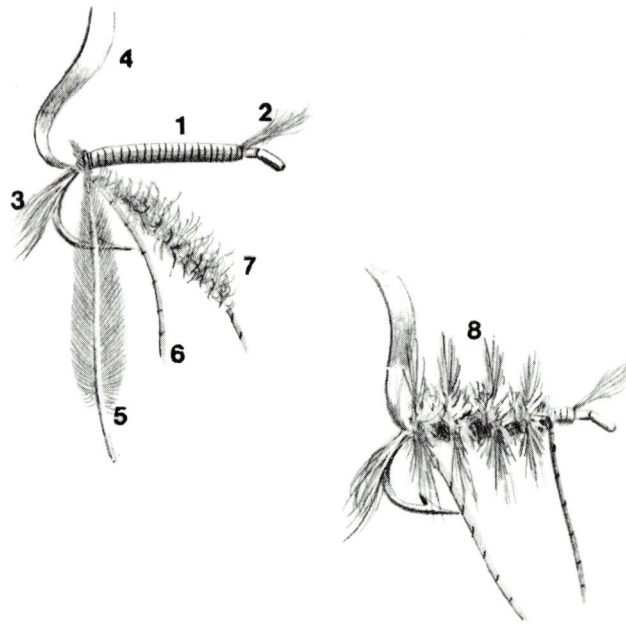

A really excellent shrimp pattern for which I have a high regard. Some while ago the man who devised this particular pattern, Taff Price, sent me an example along with the dressing instructions, wherein he stated that some three years ago while studying the shrimps in his experimental tank he noted that after shedding its old skin a shrimp was very much paler in overall hue than its companions and seem to be picked out by the trout for quick despatch. This pattern is based on the natural in that unhappy state.

The hook size is between 10 and 14 and it can be lead weighted if desired. The tying silk, 1, of a compatible colour, is started a small distance behind the eye, at which point tie in a short bunch of golden-olive cock hackle fibres, 2.

Take the silk past the bend of the hook and tie in a tail of golden-olive cock hackle fibres sloping well down the

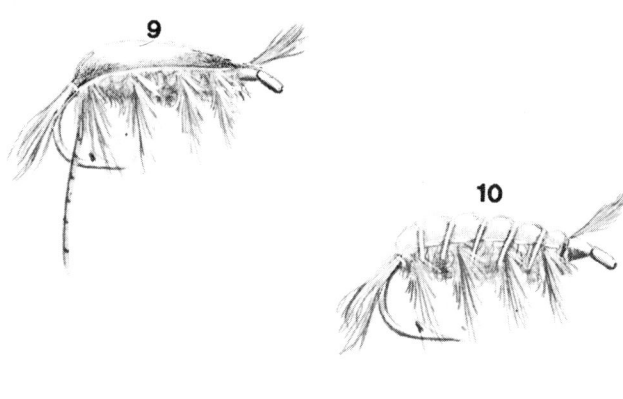

bend, 3, followed by a length of light rubber latex, 4.

Take a finely-fibred golden-olive cock hackle and tie in by the tip, 5, and then secure a length of yellow terylene ribbing thread, 6.

Wax the tying silk and dub carefully with well mixed golden-olive seal's fur, 7. Wind the silk and the dubbing up the shank to form the body, followed by carefully spaced turns of the hackle. Secure the hackle and remove the waste end, also excess dubbing, 8.

Stretch the latex over the back of the body, dividing the hackle fibres on either side, and secure at the head with the tying silk, 9. Now take the yellow terylene ribbing thread and wind in open turns over all, 10. Secure and cut off waste. Complete the pattern with a tapered and varnished whip-finish.

Fish the pattern well sunk and close to weed beds.

American Gold Ribbed Hare's Ear

Collyer

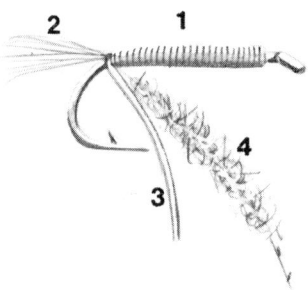

While David Collyer makes no claim to have invented this variation on the traditional Gold Ribbed Hare's Ear he would seem to have been responsible for its popularity as a capital stillwater pattern in this country.

As the name implies it is the American version of the British pattern and in its trans-Atlantic guise it has been a regular standby of the American flyfisher for many years, though it first came to Collyer's notice in 1968 via the publication *Field and Stream*. Collyer tied up a few examples for his next trip to Darwell reservoir where in very short order he took the limit of six trout. His 'battle order' that day was a forward taper sink tip line with the American pattern on the leader tip going well down to the Darwell trout that on that day were lying deep. However the pattern is a firm favourite at any depth, for on its next outing in the hands of Dave Collyer he fished it in the surface film at Chew during a buzzer rise and took trout in no uncertain manner. I have fished the fly on streams and rivers as a substitute for the standard Gold Ribbed Hare's Ear and I have found it to be just as useful as its forebear.

The original American dressing called for a long-shank

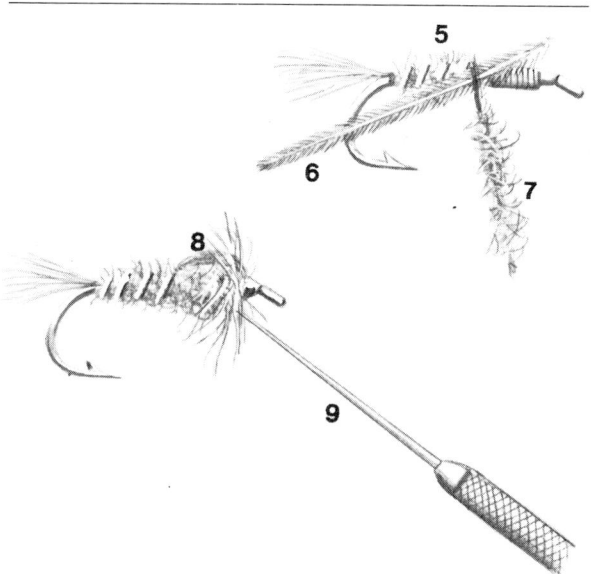

hook but Collyer has popularized it with a standard length wide-gape hook, between sizes 10 and 14. Take the black or brown tying silk down the hook shank in even turns to the bend, 1, and there tie in a bunch, a dozen or so, of hare's fur fibres for the tail, 2. Now tie in a length of gold oval tinsel for the ribbing material, 3. Re-wax the tying silk and carefully dub it with hare's body fur, 4, and wind into a gently tapered abdomen shape, followed by wide even turns of ribbing, 5.

Tie in a length of dyed black turkey tail feather fibre, 6, so that it is located on top of the hook shank. Re-dub the tying silk with Hare's fur of longer fibre than that used for the abdomen, 7. Wind this round the hook shank to form a pronounced thorax. Now bring the turkey tail fibres over the top of the thorax, doubling and re-doubling as required, to form the wing cases, 8.

Having completed the fly with a whip-finished tapered head take a dubbing needle and tease out from the thorax a few long strands of fur to make a soft hackle.

Fished at any depth in still or running water this is a very good pattern.

Pheasant Tail

Cove

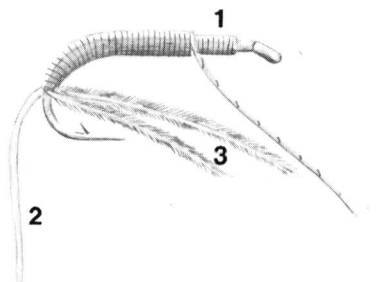

When Brian Clarke, author of the best selling *The Pursuit of Stillwater Trout*, 1975, writes that he never travels far without a selection of this Pheasant Tail creation in various sizes then we should all sit up and take notice, for Clarke is a highly skilled practitioner in the taking of still-water trout.

This particular Pheasant Tail was devised by Arthur Cove of Wellingborough, Northamptonshire, himself a most knowledgeable angler and an authority on stillwater fly patterns. While this artificial bears some resemblance to the Frank Sawyer Pheasant Tail Nymph its general configuration is more in keeping with the midge pupa. It was designed to be fished on a long leader and a floating line and in fact I have found it useful to wrap an underbody of copper wire, or similar, to ensure that the pattern sinks well on the initial cast.

Brian Clarke tells me that he has caught vast numbers of trout with this artificial when drawing it slowly across the bottom of a lake. It is also a great success fished in mid-water, Clarke being of the opinion that its success may lie

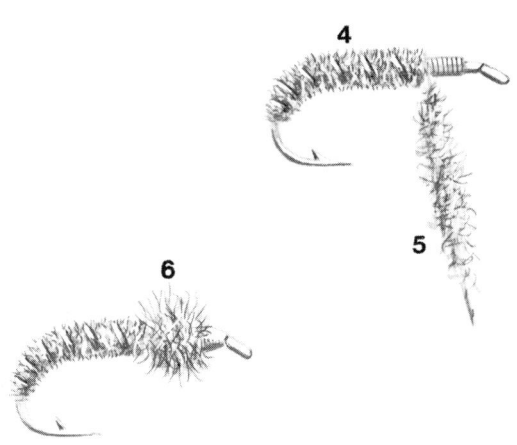

in its exaggerated form of a midge pupa.

The pattern is very easy to tie, though care should be taken to achieve the correct shape. The hook size is generally between 8 and 6. Take the brown tying silk, 1, down the hook shank in close even turns to a point half way around the bend. There tie in a length of fine copper wire that will be used for the ribbing, 2. Now tie in strands of cock pheasant centre tail feather fibres, 3. Wind the silk back up the hook shank to a point approximating to the rear of a thorax.

In very even turns wind the pheasant tail fibres up to the silk. These can be twisted together if you find it easier to wind them that way. Secure with the tying silk and cut off the waste. Now take the copper wire in even turns over the fibres, creating the ribbed effect, 4. Remove waste wire.

Re-wax the silk and dub with rabbit or hare's body fur, 5. Wind the dubbed silk backwards and forwards, creating a pronounced thorax shape, 6. Complete the fly with a neat whip-finish, well varnished.

Barney Google

Walker

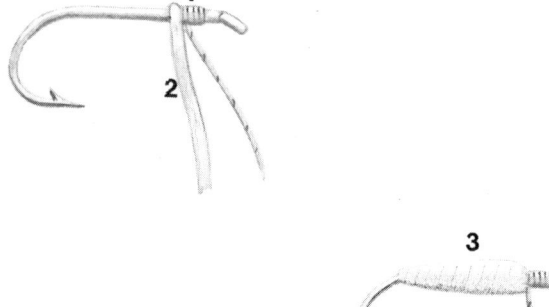

This oddly named pattern is Richard Walker's tying of a member of the *Chaoborus* family, commonly known as the phantom larvae because of their almost translucent appearance. A pattern of the same species by David Collyer has already been described in this book, and it is interesting to compare Collyer's dressing with this one.

As Walker rightly states in his book *Fly Dressing Innovations*, 1974, the phantom larvae are not easy to imitate, for they are in the main transparent, with only a large bulbous thorax that turns pale brown, sepia or orange as pupation progresses.

Walker's method of imitating the latter is to use two small transparent red beads tied in at the front of the body. He comments that while the end result looks very gaudy in the flybox when the pattern has sunk six feet or so into the water the red beads appear as a dark sepia colour.

The fly should be fished on a floating line with a long leader and allowed to sink well down before being retrieved in a series of small jerks.

The pattern is tied on a size 14 hook, though its inventor suggests a size 12 with a shortened body if one is fishing a

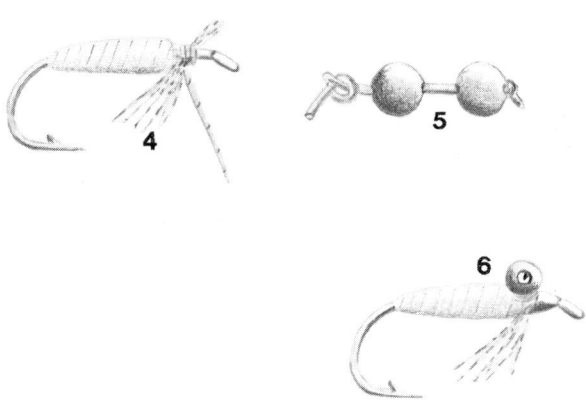

location that holds larger than average fish. Start the tying silk from a short distance behind the eye, 1, and wind for four or five turns. There tie in a narrow strip of clear polythene, 2. Wind the polythene back and forth along the hook shank to form a thin body, 3. One can experiment with white painted or silvered hook shanks for alternative effects.

Tie in a small beard-hackle made from a few fibres of grey speckled mallard, partridge or widgeon feather, 4. We now come to the interesting part. Two small transparent balls or beads have to be joined together. Walker's original dressing had them joined by a short length of tying silk, threading first one and then another onto the silk, and restraining them with a knot, 5. The inventor's later dressings used fine copper wire in place of the tying silk.

Having secured the beads to the copper wire place them at right angles to the hook shank just in front of the body and secure with figure-of-eight turns of tying silk, 6, continuing the silk forward to form a neatly-tapered head which is then whip-finished and varnished.

Early Olive Nymph

Bucknall

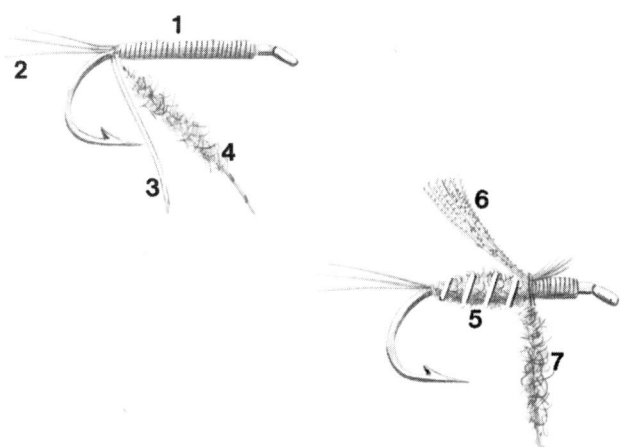

Geoffrey Bucknall is a personality well known to the fly-fishing public via his books and his many articles in the angling press. His name will always be linked to his rightly famous Footballer series of flies for still water but his experience is not confined to the reservoirs of this country, for he is equally skilled on flowing waters from the southern chalk streams to the rivers of the north.

Bucknall's dressing of the Early Olive Nymph was developed from one of these northern rivers, the Border Esk, and it is a pattern that he considers to be one of the most effective he has ever devised. Certainly it looks the part and I would draw your attention to his method of imitating the legs of the original nymph. Bucknall does not use a wound hackle to imitate the legs but puts to good use the waste ends of the thorax material, a method that I have adopted on many dressings.

All his patterns are developed from life. The naturals he collects are photographed and then enlarged many times so that he can study the overall shape before translating the subject into a fur and feather replica. Though you may not have the required photographic equipment I would urge

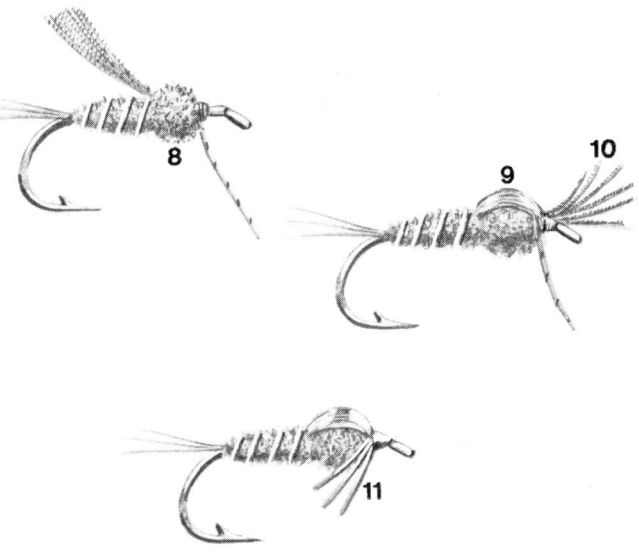

you to use a magnifying glass and study the insect you wish to represent rather than slavishly follow existing patterns.

The hook size is 12, though I find size 14 equally effective. Take the dark-grey tying silk, 1, down the shank in close even turns to the bend at which point tie in three fibres from a dark-olive cock hackle to represent the tails, 2. Now tie in a length of fine gold wire, 3, and re-wax the tying silk before dubbing it with mole or water-rat fur, 4.

Wind the dubbed silk up the hook shank to form the abdomen, followed by the gold wire as the rib, 5. Remove the waste end of the rib and strip off the excess dubbing.

Tie in a bunch of water-hen feather fibres, 6, and re-dub the silk with dark-olive seal's fur, 7. Wind the seal's fur round the hook shank to form the thorax, 8. Remove excess dubbing.

Bring the water-hen fibres over the top of the thorax to form the wing cases, 9, and secure with a half-hitch. Equally divide the wing case fibres on either side of the thorax, 10, and draw back to either side of the thorax to form the legs, 11. Secure with a firmly-wound whip-finish and varnish the head.

Mayfly Nymph

Bradbury

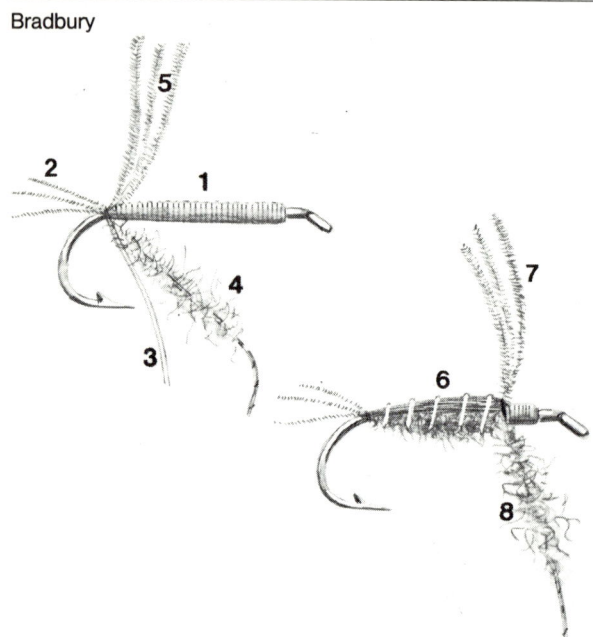

Derek Bradbury is one of the most inventive and observant flytyer/entomologists on the contemporary angling scene and I have full confidence in any pattern that he has designed, especially his mayfly dressings, both floating and nymphal. Having the luck to fish a stream that still boasts a hatch of mayfly I can certainly vouch for the efficacy of his carefully observed patterns, while this nymph has the advantage of being not only very effective but also very durable.

You will note that Bradbury imitates the colour variation between the dorsum and ventor sections of the body in a most simple, though effective, manner.

The tying of this mayfly nymph is not difficult, but it does require care to achieve a good looking artificial.

The hook is generally a long-shank size 10. The tying silk of an olive colour, 1, is wound down the shank in close even turns to the bend of the hook. Then tie in three fibres from the centre tail feather of a cock pheasant, 2. Follow

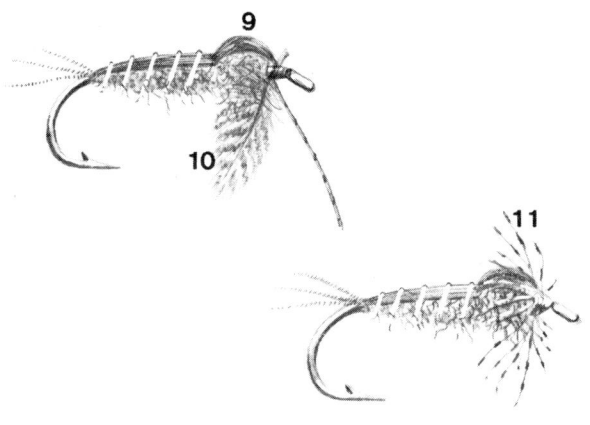

this with a length of oval gold tinsel, 3. Now dub the tying silk with yellow-olive seal's fur, 4. Lift the waste ends of the tail fibres, 5, out of the way and carefully wind the dubbed silk to form a tapered body. Bring the waste ends of the tail fibres over the top of the body and bind in place with the gold oval tinsel, 6. Do not cut off the waste ends of pheasant herls, 7.

Re-wax the silk and dub again with a further quantity of yellow-olive seal's fur, 8, and wind onto the hook shank to form a pronounced thorax. Bring the pheasant tail herls over the thorax and tie down. Double and re-double as necessary to form the wing cases, 9.

Now tie in a brown partridge breast feather, 10. Wind one turn only of this hackle round the hook shank in front of the thorax, 11, and complete the fly with a tapered, varnished head.

Truly a great pattern and one that has proved its worth on waters that have never known the mayfly!

No-name Nymph

Overfield

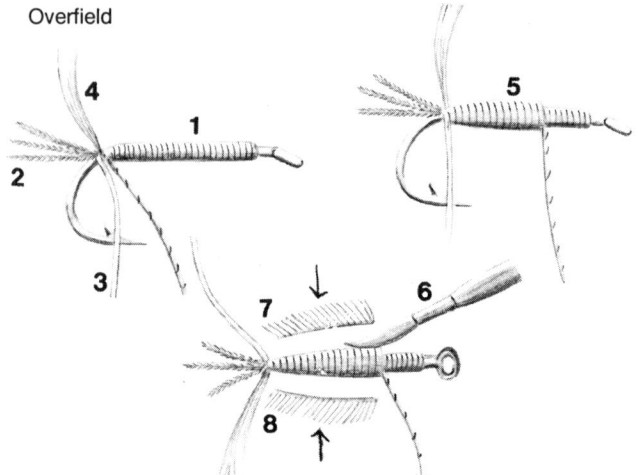

I think that all author-illustrators should be allowed one little whimsy after the hard slog . . . yes, it is, whatever you may think . . . of producing fifty water colour and pencil drawings for such a book as this. A number of times when I was wielding a number 00 paintbrush I wished that I had been flexing an eight foot fly rod! However this first volume of the series is now complete and to celebrate the occasion I will give you a pattern of my own that has no name, that will tax your patience and your ingenuity in its fabrication and will not be very durable, but which is good, very good, if you will try it and cast to a nymphing trout of the streams.

When I get a little bored with tying what we may choose to call standard patterns I busk up a few of these, mainly for the material manipulation exercise involved. Try it. You may well swear at me but you will know a lot more about the dextrous use of the materials and tools at the end of the session.

You will note that I have not stipulated any specific colours, for you can adapt the pattern to any number of naturals.

My usual hook size is between 14 and 12. Take the tying silk, 1, down to the bend in close even turns and there tie in three tail whisks, 2, of pheasant or condor herl. Follow

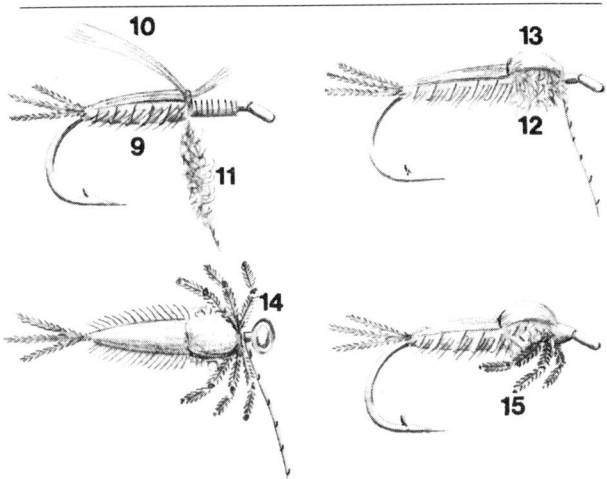

these herls with a length of gold or silver wire, 3, and on top of the hook shank tie a length of PVC, colour to suit, 4, that is cut on a taper towards the tail.

Wind the silk in close even turns to create the shape of the abdomen, 5. Now take two cock hackles of the appropriate colour and strip off the fibres from one side, trimming the remaining side with sharp scissors. If you wish leave the latter operation until the hackle stalk is attached to the body. With a fine brush carefully put a line of adhesive along the silk wound abdomen, 6, and then place the prepared hackle stalks either side of the body, 7 and 8.

Carefully wind the ribbing wire up the abdomen in spaced turns, trapping the hackle stalks en route. Now stretch the PVC over the back of the abdomen and tie in, 9. Tie in your choice of material for the wing cases, 10, and then dub the tying silk with some appropriate fur, 11.

Wind the dubbed silk to create a pronounced thorax, 12, and bring the wing case material over all and tie in, 13, removing the waste ends. Now tie in three crossed lengths of pheasant tail or heron herl to simulate the legs and tie in with a figure-of-eight whip, 14. Complete the fly with a whip-finish, 15.

It is finicky, but great fun. Go to it!

Index

*Printed in Great Britain
by Ebenezer Baylis and Son, Limited,
The Trinity Press,
Worcester, and London*